Approximating Distances Between World Capitals

Applying the Spherical Law of Cosines for Sides

Dale Johanson

Cover photography and proof illustrations
by Seth Johanson

Curve Beneath the Cloud LLC
ISBN-10: 0-9978303-0-1
ISBN-13: 978-0-9978303-0-9

To Jamie

Contents

Preface

Spherical trigonometry is a truly fascinating subject that declined in popularity in the middle of the twentieth century and is rarely taught today. In this activity book, you will use one of the laws from that subject to calculate the approximate distance between two capital cities.

Chapter one, "Laying the Foundation", begins with a short list of topics from plane trigonometry. You might consider checking out a trigonometry text at your local library and reviewing examples of each of the items from the list. The chapter then continues with a proof of thc spherical law of cosines for sides, and concludes with additional information and illustrations to help you get started in performing your calculations.

Chapters two through eleven are devoted to calculating distances between capitals within a region or continent. In some cases, more than one capital may exist. For example, a country may have a legislative capital as well as an administrative capital. If that is indeed the case, I will reference the latitude and longitude of the administrative capital. There are a total of 18 examples and 98 problems with solutions in the book. I hope you enjoy calculating the distances between capitals using the spherical law of cosines for sides!

Dale Johanson
September 14, 2017

Chapter 1

Laying the Foundation

Before you embark on your journey around the world via the spherical law of cosines for sides you may want to study and/or review a few key items from plane trigonometry. The main items include the

1. sine function
2. cosine function
3. inverse cosine function
4. arc length formula
5. Pythagorean Theorem
6. law of cosines.

In addition to those items from plane trigonometry, other items of importance that you may want to study and/or review include

1. the equator
2. the prime meridian
3. latitude
4. longitude

In the problems throughout the book, a longitude will either be referenced in degrees west of the prime meridian or in degrees east of the prime meridian.

You can use either a scientific or graphing calculator. Make sure the calculator is in degree mode, and float mode. Basically, we don't want the calculator to round to a fixed number of decimal places.

After you've completed the tasks outlined above you can either read through the proof of the spherical law of cosines for sides which starts on the next page, or you can skip to page 14 where all the steps involved in calculating the approximate distance between two capitals are explained.

Imagine a plane intersecting a sphere at single point A. Extend a line segment from the center of the sphere, A_1, outward intersecting the sphere at point B and then intersecting the plane at point B_1. Next, extend another line segment from the point, A_1, outward intersecting the sphere at point C and then intersecting the plane at point C_1. Finally, extend another line segment from the point, A_1, outward intersecting both the sphere and the plane at point A. Unlike the sides of the triangle in a plane, the sides of the spherical triangle formed by the arcs a, b, and c, respectively, are themselves angles. The given sides a, b, and c also represent arcs on great circles. A great circle is the circle formed by the intersection of a plane passing through the center of a sphere and the sphere itself.

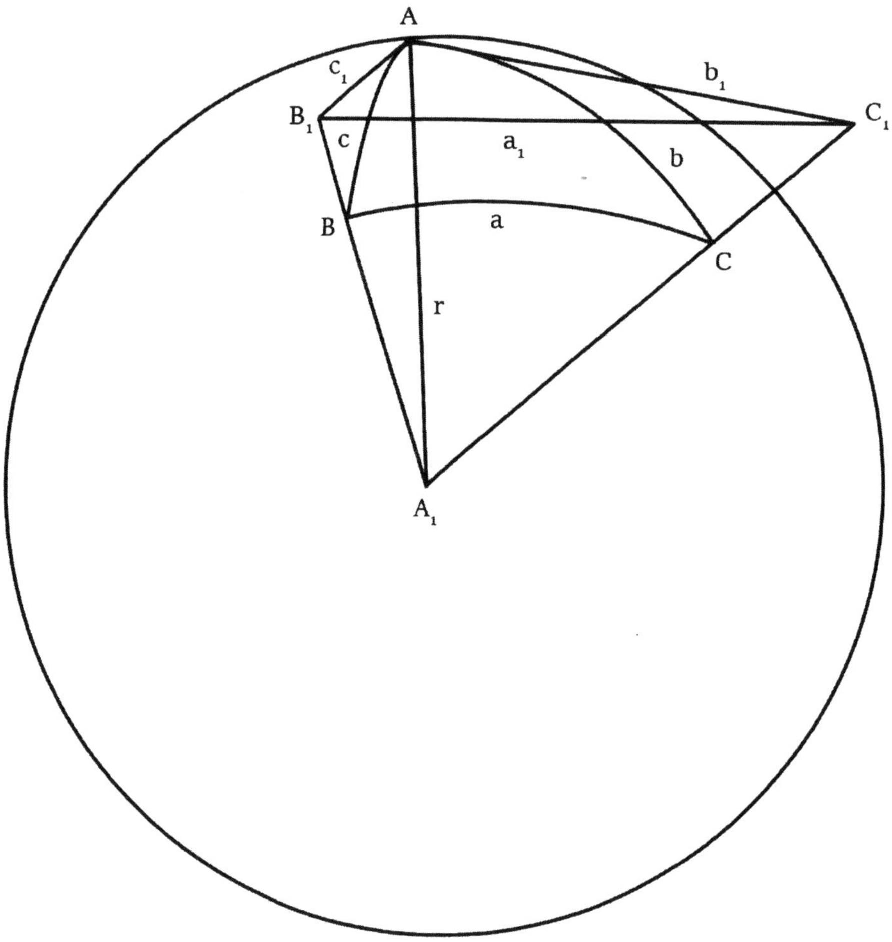

Figure # 1

As we proceed through the steps, we consider four triangles created in Figure # 1.

We first examine the triangle,

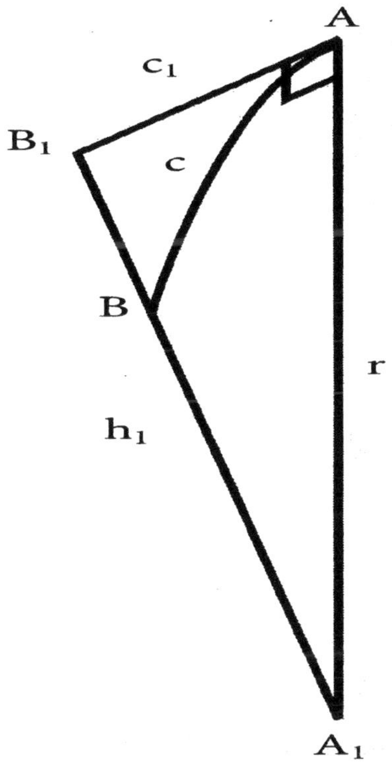

Figure # 2

Let h_1 represent the hypotenuse of the right triangle above. It follows that,

$\cos c = \frac{r}{h_1} \Rightarrow h_1 \cos c = r \Rightarrow h_1 = \frac{r}{\cos c}$ and $\sin c = \frac{c_1}{h_1} \Rightarrow h_1 \sin c = c_1 \Rightarrow c_1 = \left(\frac{r}{\cos c}\right) \sin c$

$\Rightarrow c_1 = \frac{r \sin c}{\cos c}$ (equation # 1)

Second, we examine the triangle,

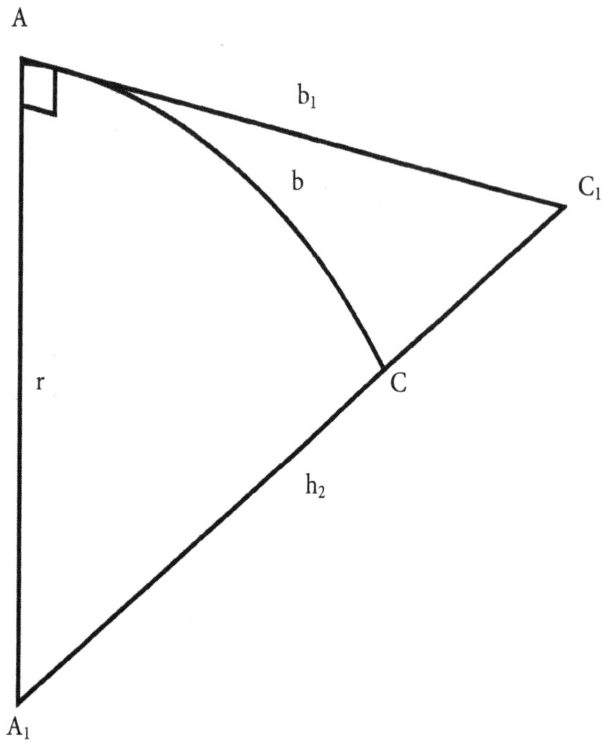

Figure # 3

Let h_2 represent the hypotenuse of the right triangle above. It follows that,

$\cos b = \frac{r}{h_2} \Rightarrow h_2 \cos b = r \Rightarrow h_2 = \frac{r}{\cos b}$ and $\sin b = \frac{b_1}{h_2} \Rightarrow h_2 \sin b = b_1 \Rightarrow b_1 = \left(\frac{r}{\cos b}\right) \sin b$

$\Rightarrow b_1 = \frac{r \sin b}{\cos b}$ (equation # 2)

Third, we consider the triangle on the plane.

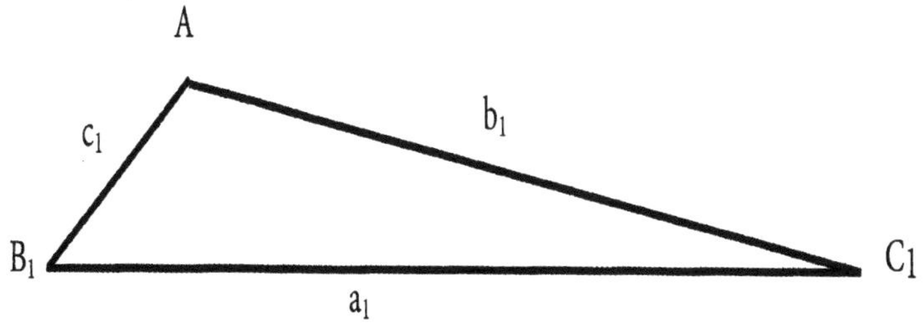

Figure # 4

Using the law of cosines on a plane,

$$a_1{}^2 = b_1{}^2 + c_1{}^2 - 2b_1c_1 \cos A$$

$$= \left(\frac{r \sin b}{\cos b}\right)^2 + \left(\frac{r \sin c}{\cos c}\right)^2 - 2\left(\frac{r \sin b}{\cos b}\right)\left(\frac{r \sin c}{\cos c}\right) \cos A \quad \text{using equations \# 1 and \# 2}$$

$$= \frac{r^2 \sin^2 b}{\cos^2 b} + \frac{r^2 \sin^2 c}{\cos^2 c} - \frac{2\, r^2 \sin b \sin c \cos A}{\cos b \cos c} \qquad \text{(equation \# 3)}$$

Fourth, we consider the triangle,

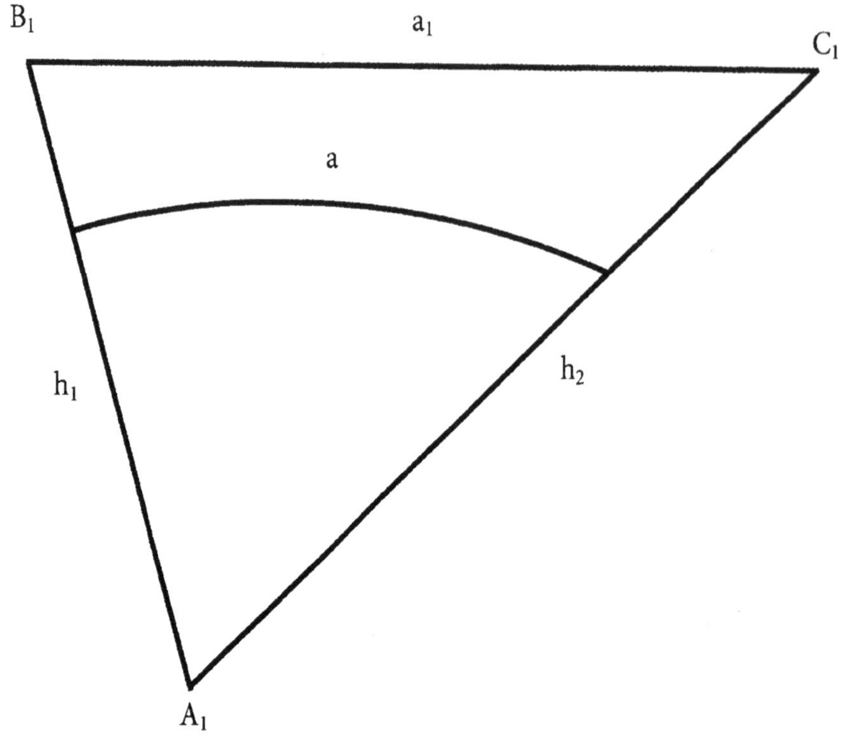

Figure # 5

Once again, using the law cosines on the plane, ${a_1}^2 = {h_2}^2 + {h_1}^2 - 2h_2h_1 \cos a$

$$= \left(\frac{r}{\cos b}\right)^2 + \left(\frac{r}{\cos c}\right)^2 - 2\left(\frac{r}{\cos b}\right)\left(\frac{r}{\cos c}\right)\cos a$$

$$= \frac{r^2}{\cos^2 b} + \frac{r^2}{\cos^2 c} - \frac{2\,r^2 \cos a}{\cos b \cos c} \qquad \text{(equation \# 4)}$$

We now set the result of equation # 3 equal to the result of equation # 4,

$$\frac{r^2 \sin^2 b}{\cos^2 b} + \frac{r^2 \sin^2 c}{\cos^2 c} - \frac{2\,r^2 \sin b \sin c \cos A}{\cos b \cos c} = \frac{r^2}{\cos^2 b} + \frac{r^2}{\cos^2 c} - \frac{2\,r^2 \cos a}{\cos b \cos c}$$

After multiplying both sides of the newly formed equation by $\cos^2 b \cos^2 c$ we obtain,

$$r^2\sin^2 b \cos^2 c + r^2\sin^2 c \cos^2 b - 2r^2 \sin b \sin c \cos b \cos c \cos A$$
$$= r^2\cos^2 c + r^2\cos^2 b - 2r^2 \cos b \cos c \cos a$$

Next, we divide both sides of the equation by r^2 to obtain

$$\sin^2 b \cos^2 c + \sin^2 c \cos^2 b - 2 \sin b \sin c \cos b \cos c \cos A$$
$$= \cos^2 c + \cos^2 b - 2 \cos b \cos c \cos a$$

Next, we isolate the term containing $\cos a$ on the left side of the equation,

$$2 \cos b \cos c \cos a + \sin^2 b \cos^2 c + \sin^2 c \cos^2 b - 2 \sin b \sin c \cos b \cos c \cos A$$
$$= \cos^2 c + \cos^2 b$$

$$2 \cos b \cos c \cos a$$
$$= \cos^2 c + \cos^2 b - \sin^2 b \cos^2 c - \sin^2 c \cos^2 b + 2 \sin b \sin c \cos b \cos c \cos A$$

$$2 \cos b \cos c \cos a$$

$$= \cos^2 c - \sin^2 b \cos^2 c + \cos^2 b - \sin^2 c \cos^2 b + 2 \sin b \sin c \cos b \cos c \cos A$$

$$2 \cos b \cos c \cos a = \cos^2 c(1 - \sin^2 b) + \cos^2 b(1 - \sin^2 c) + 2 \sin b \sin c \cos b \cos c \cos A$$

We now utilize the Pythagorean identity, $\sin^2\theta + \cos^2\theta = 1$, as follows:

$$2 \cos b \cos c \cos a = \cos^2 c \cos^2 b + \cos^2 b \cos^2 c + 2 \sin b \sin c \cos b \cos c \cos A$$

$$2 \cos b \cos c \cos a = 2\cos^2 b \cos^2 c + 2 \sin b \sin c \cos b \cos c \cos A$$

Finally, we divide both sides of the equation by $2 \cos b \cos c$ to obtain

$$\cos a = \cos b \cos c + \sin b \sin c \cos A$$

which is the law of cosines for sides in spherical trigonometry.

A Closer Look at the First Example

In first example of Chapter two, the approximate distance between Malabo, Republic of Equatorial Guinea, and Pretoria, South Africa is calculated. First, the geographic coordinates of both capitals are converted into a degree decimal format where each coordinate is rounded to four decimal places:

Malabo, Republic of Equatorial Guinea is located at $3^o45'\ \mathrm{N} = \left(3 + {}^{45}/_{60}\right)^o\ \mathrm{N} = 3.7500^o\ \mathrm{N}$,

$8^o47'\ \mathrm{E} = \left(8 + {}^{47}/_{60}\right)^o\ \mathrm{E} \approx 8.7833^o\ \mathrm{E}$.

Pretoria, South Africa is located at $25^o42'\ \mathrm{S} = \left(25 + {}^{42}/_{60}\right)^o\ \mathrm{S} = 25.7000^o\ \mathrm{S}$,

$28^o13'\ \mathrm{E} = \left(28 + {}^{13}/_{60}\right)^o\ \mathrm{E} \approx 28.2167^o\ \mathrm{E}$.

To create a spherical triangle, we measure the angle of the side that extends from the North Pole to Malabo. Since Malabo lies in the northern hemisphere, we subtract 3.7500^o from 90°. Next, to find the measure of the side that extends from the North Pole to Pretoria, we add 25.7000^o to 90° since Pretoria lies in the southern hemisphere. It follows that the two sides of the spherical triangle are $90^o - 3.7500^o = 86.2500^o$ and $90^o + 25.7000^o = 115.7000^o$, respectively. The angle between the two sides at the North Pole is $28.2167^o - 8.7833^o = 19.4334^o$. See the illustration on page 26.

Next, we then use the spherical law of cosines to determine the cosine of side n, and then apply the inverse cosine function to calculate the measure of angle n in degrees:

$$\cos \mathrm{n} = \cos(86.2500^o)\cos(115.7000^o) + \sin(86.2500^o)\ \sin(115.7000^o)\cos(19.4334^o)$$

$$\cos \mathrm{n} \approx 0.819559619888 \ \Rightarrow\ \mathrm{n} \approx \cos^{-1}(0.819559619888) \ \Rightarrow\ \mathrm{n} \approx 34.959265717^o.$$

In our final step, we convert the measure of angle n from degrees to radians, and apply the arc length formula to approximate distance between the two capitals,

$$(34.959265717^o)\left({}^{\pi}/_{180^o}\right)(3959\ \text{miles})$$

$$\approx (0.610154290841)(3959\ \text{miles})$$

$$\approx\ 2{,}416\ \text{miles}.$$

Creating the illustrations

With regards to the illustrations you will find in Chapters 2 through 11, they are not spherical triangles drawn to scale. The purpose of those hand drawn illustrations is to merely provide a visual aid throughout the calculations. To create the illustrations, I used a curve ruler to help depict their three dimensional nature.

To help you create your spherical triangles, I have provided seven illustrations on the following pages. Before examining the illustrations, note that E represents a longitude that is east of the prime meridian; W represents a longitude west of the prime meridian; N represents a latitude north of the equator and S represents a latitude south of the equator. The subscripts identify the city being referenced. For example, E_1 represents a longitude east of the prime meridian for city B, whereas, N_2 represents a latitude north of the equator for city C. If you examine the notes below each of the seven illustrations, you will see how the measure of the angle at the pole, P, as well as the length of side's b and c are calculated. Finally note, it is important to remember that in spherical trigonometry the sides of a spherical triangle can always be thought of in two ways. They are angles of a given measure, and they are sides of a given length.

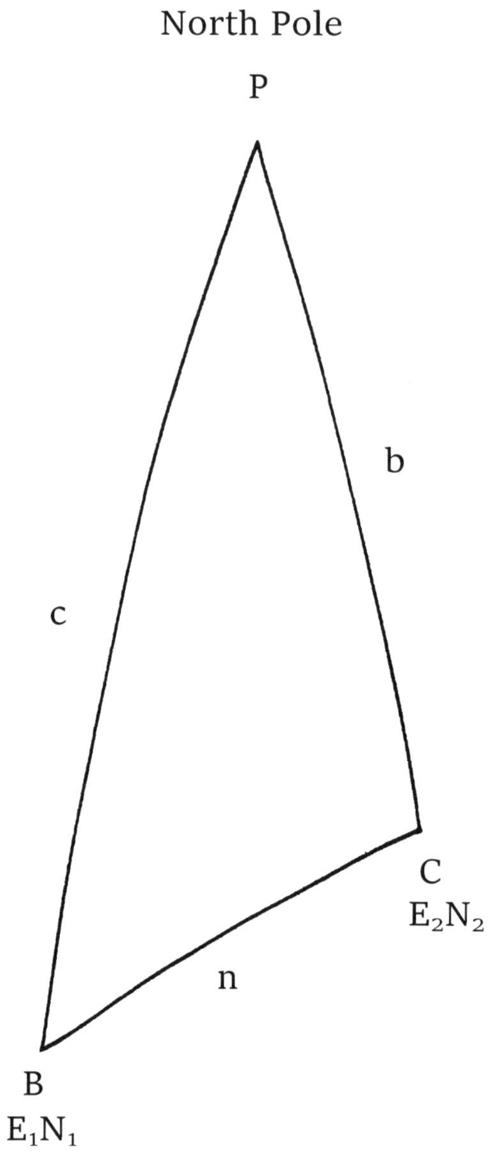

where $E_2 > E_1$

$P = E_2 - E_1$

$c = 90^o - N_1$

$b = 90^o - N_2$

Figure # 6

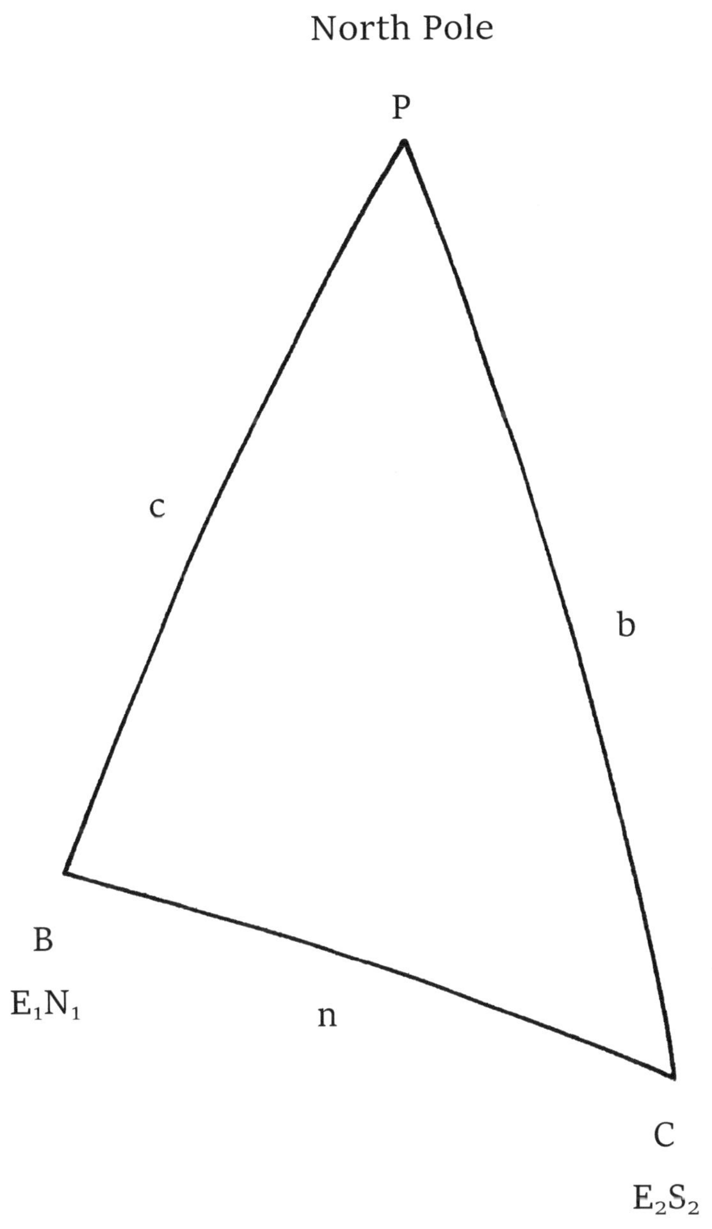

where $E_2 > E_1$

$P = E_2 - E_1$

$c = 90^o - N_1$

$b = 90^o + S_2$

Figure # 7

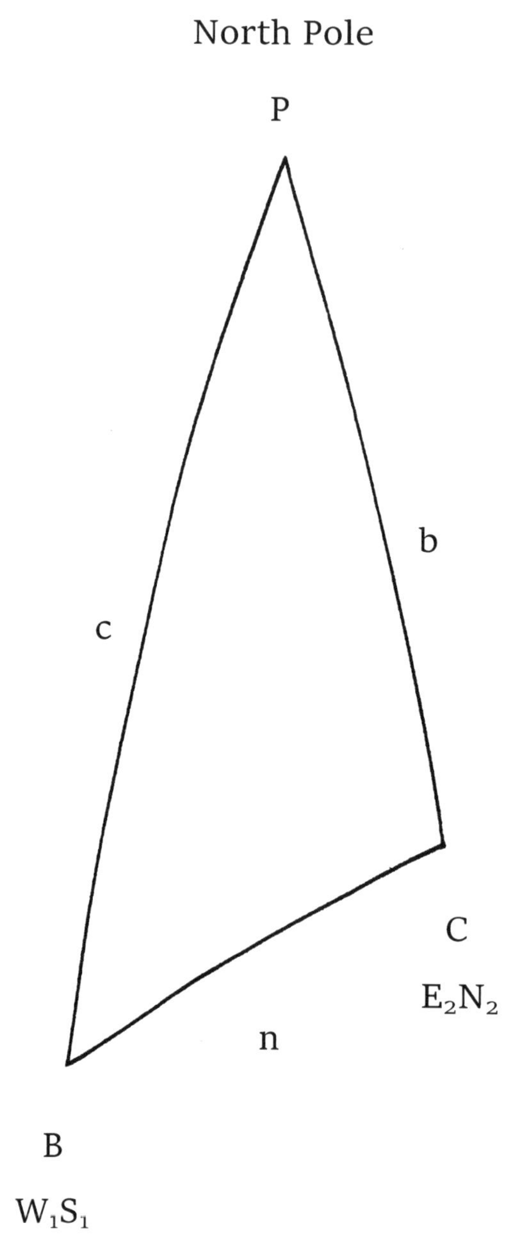

$P = W_1 + E_2$

$c = 90^o + S_1$

$b = 90^o - N_2$

Figure # 8

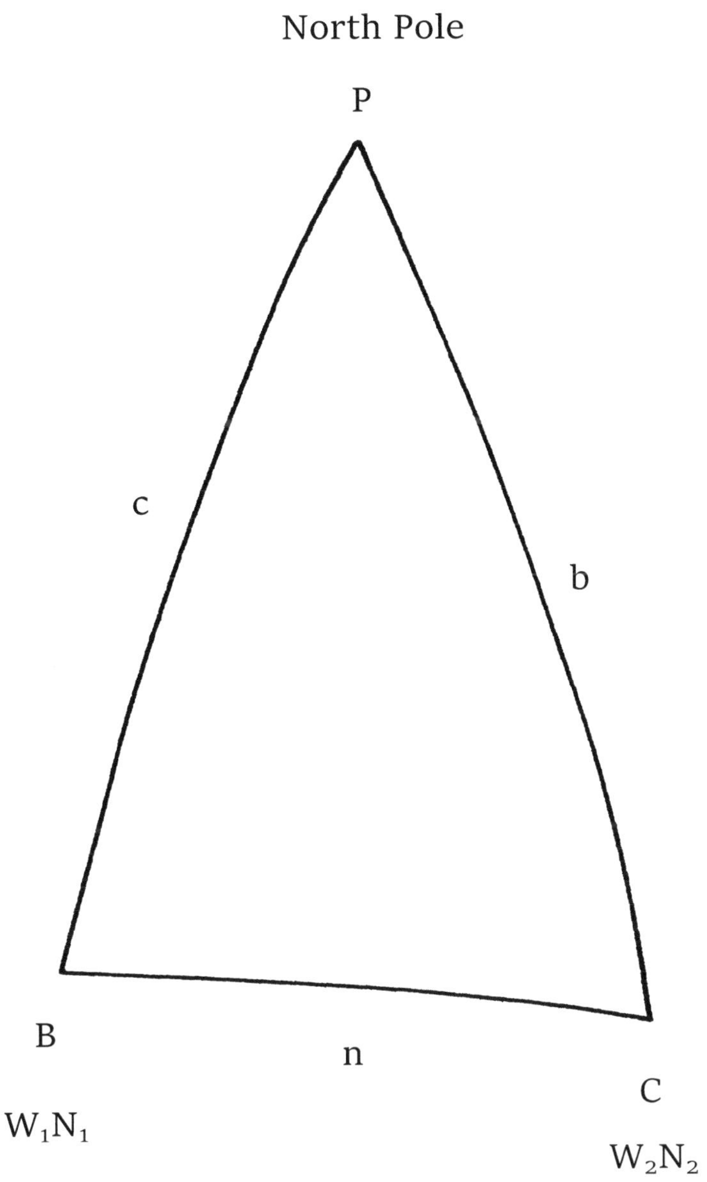

where $W_1 > W_2$

$P = W_1 - W_2$

$c = 90^o - N_1$

$b = 90^o - N_2$

Figure # 9

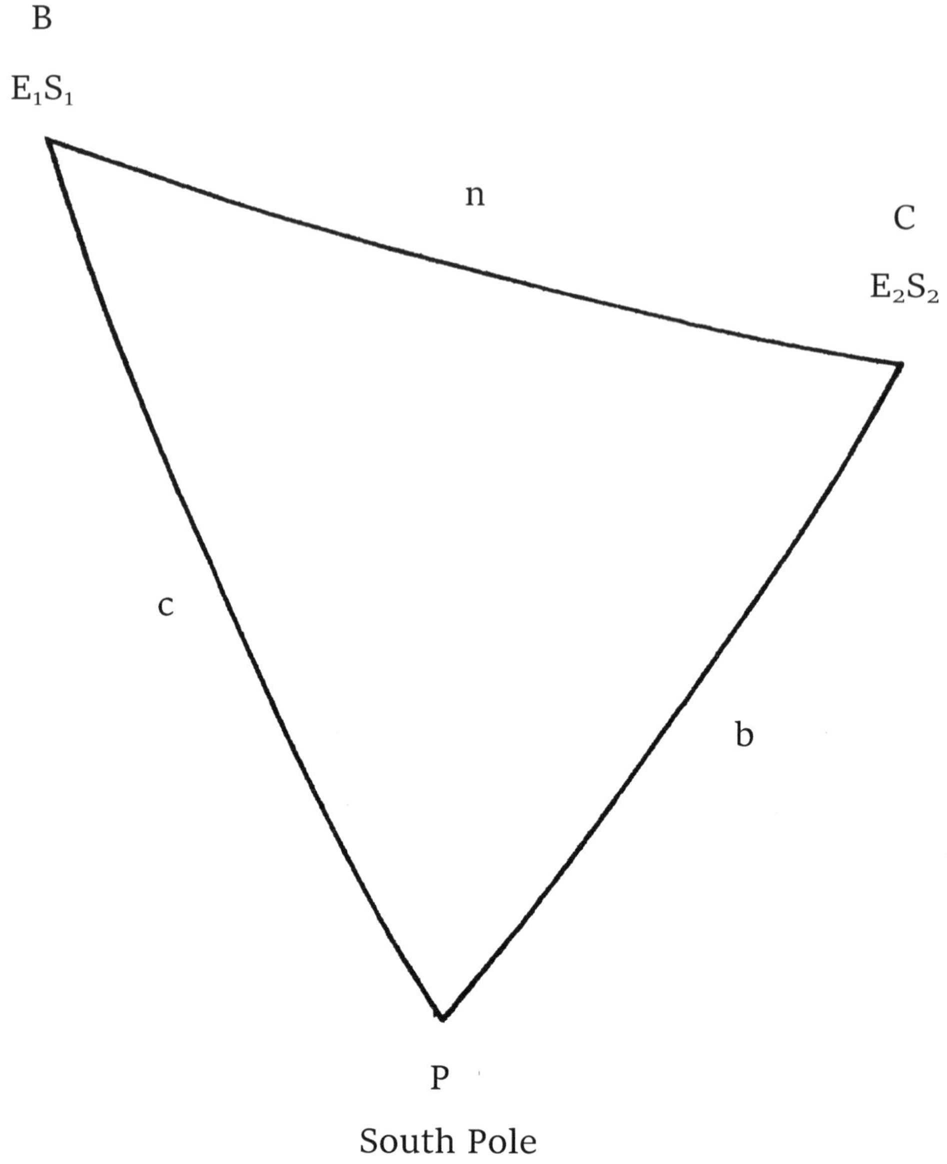

where $E_2 > E_1$

$P = E_2 - E_1$

$c = 90^\circ - S_1$

$b = 90^\circ - S_2$

Figure # 10

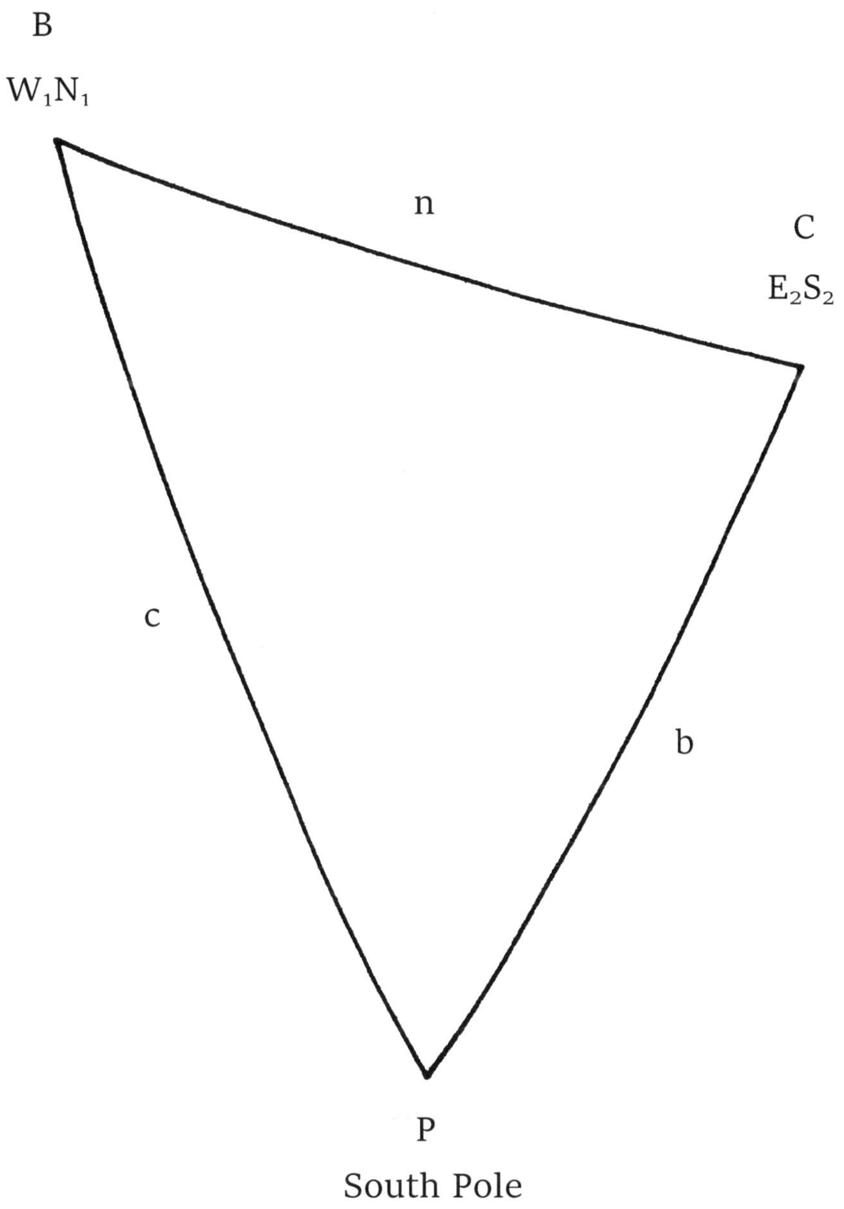

$P = W_1 + E_2$

$c = 90^\circ + N_1$

$b = 90^\circ - S_2$

Figure # 11

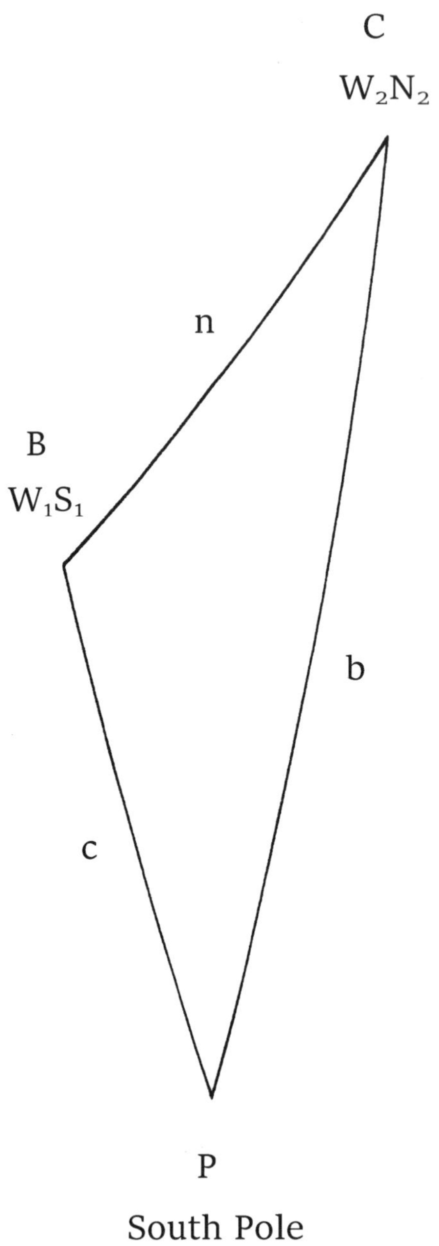

where $W_1 > W_2$

$P = W_1 - W_2$

$c = 90^o - S_1$

$b = 90^o + N_2$

Figure # 12

Chapter 2

Africa

Example 1

Malabo, Republic of Equatorial Guinea is located at

$3^o45'\ \mathrm{N} = \left(3 + {}^{45}/_{60}\right)^o\ \mathrm{N} = 3.7500^o\ \mathrm{N}\,,\ 8^o47'\ \mathrm{E} = \left(8 + {}^{47}/_{60}\right)^o\ \mathrm{E} \approx 8.7833^o\ \mathrm{E}\,.$

Pretoria, South Africa is located at

$25^o42'\ \mathrm{S} = \left(25 + {}^{42}/_{60}\right)^o\ \mathrm{S} = 25.7000^o\ \mathrm{S}\ ,\ 28^o13'\ \mathrm{E} = \left(28 + {}^{13}/_{60}\right)^o\ \mathrm{E} \approx 28.2167^o\ \mathrm{E}\,.$

Two sides of the spherical triangle are $90^o - 3.7500^o = 86.2500^o$ and $90^o + 25.7000^o = 115.7000^o$, respectively. The angle at the North Pole is $28.2167^o - 8.7833^o = 19.4334^o$.

See the illustration on the next page.

It follows that

$\cos n = \cos(86.2500^o)\cos(115.7000^o) + \sin(86.2500^o)\ \sin(115.7000^o)\cos(19.4334^o)$

$\cos n \approx 0.819559619888 \ \Rightarrow\ n \approx \cos^{-1}(0.819559619888)\ \Rightarrow\ n \approx 34.959265717^o.$

Therefore, the approximate distance between the two capital cities is

$(34.959265717^o)\left({}^{\pi}/_{180^o}\right)(3959\ \text{miles})$

$\approx (0.610154290841)(3959\ \text{miles})$

$\approx\ 2{,}416\ \text{miles}\,.$

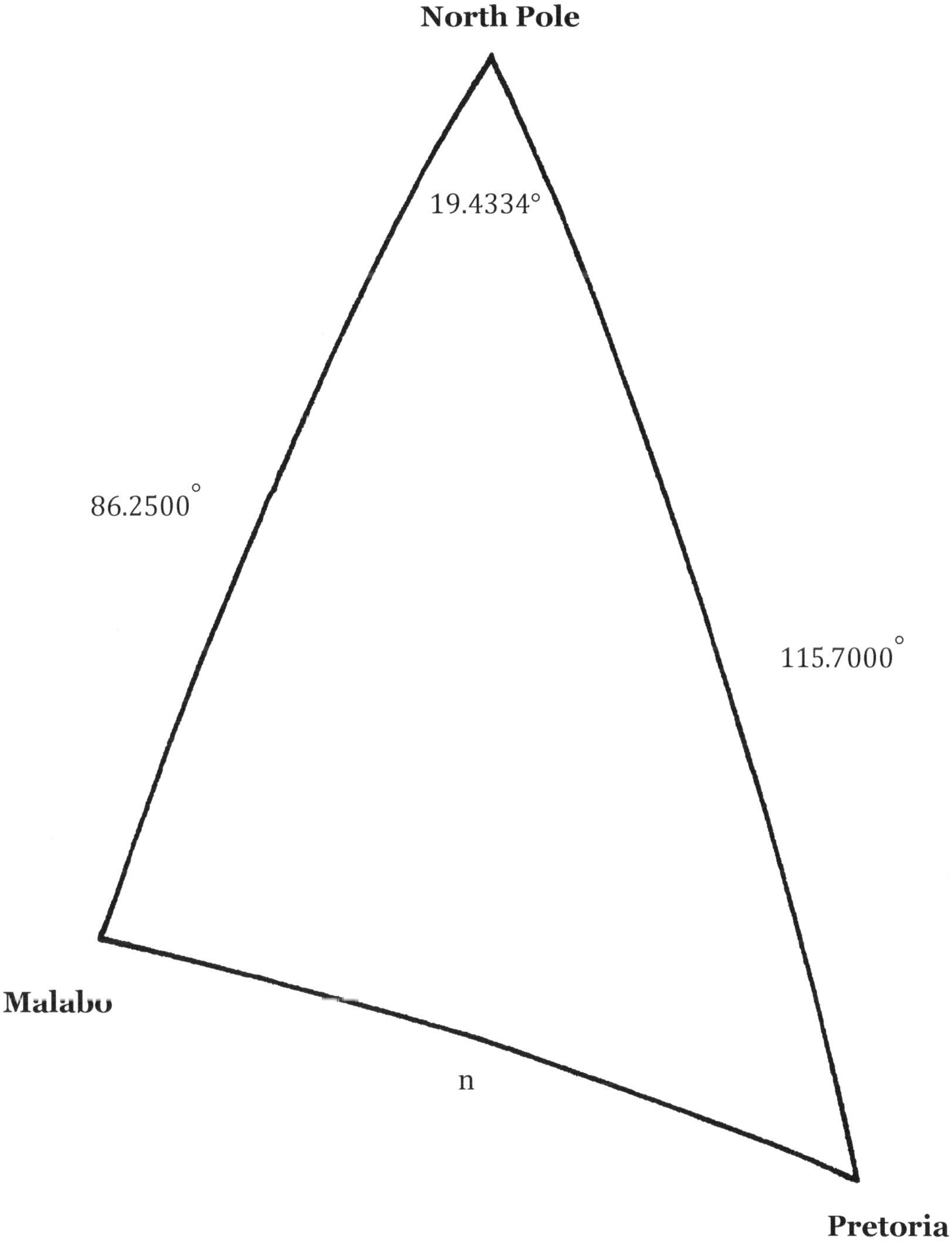
North Pole
19.4334°
86.2500°
115.7000°
Malabo
n
Pretoria

Example 2

Bissau, Guinea-Bissau is located at $11^o51'$ N $= \left(11 + {}^{51}/_{60}\right)^o$ N $= 11.8500^o$ N ,
$15^o35'$ W $= \left(15 + {}^{35}/_{60}\right)^o$ W $\approx 15.5833^o$ W .

Dar es Salaam, Tanzania is located at $6^o48'$ S $= \left(6 + {}^{48}/_{60}\right)^o$ S $\approx 6.8000^o$ S ,
$39^o17'$ E $= \left(39 + {}^{17}/_{60}\right)^o$ E $\approx 39.2833^o$ E .

Two sides of the spherical triangle are $90^o + 11.8500^o = 101.8500^o$ and
$90^o - 6.8000^o = 83.2000^o$, respectively. The angle at the South Pole is
$15.5833^o + 39.2833^o = 54.8666^o$.

See the illustration on the next page.

It follows that

$$\cos n = \cos(101.8500^o)\cos(83.2000^o) + \sin(101.8500^o)\sin(83.2000^o)\cos(54.8666^o)$$

$$\cos n \approx 0.534941505514 \quad \Rightarrow n \approx \cos^{-1}(0.534941505514\ ) \Rightarrow n \approx 57.6600554513^o.$$

Therefore, the approximate distance between the two capital cities is
$(57.6600554513^o)\left({}^{\pi}/_{180^o}\right)(3959 \text{ miles})$

$\approx (1.00635781451)(3959 \text{ miles})$.

$\approx$ 3,984 miles.

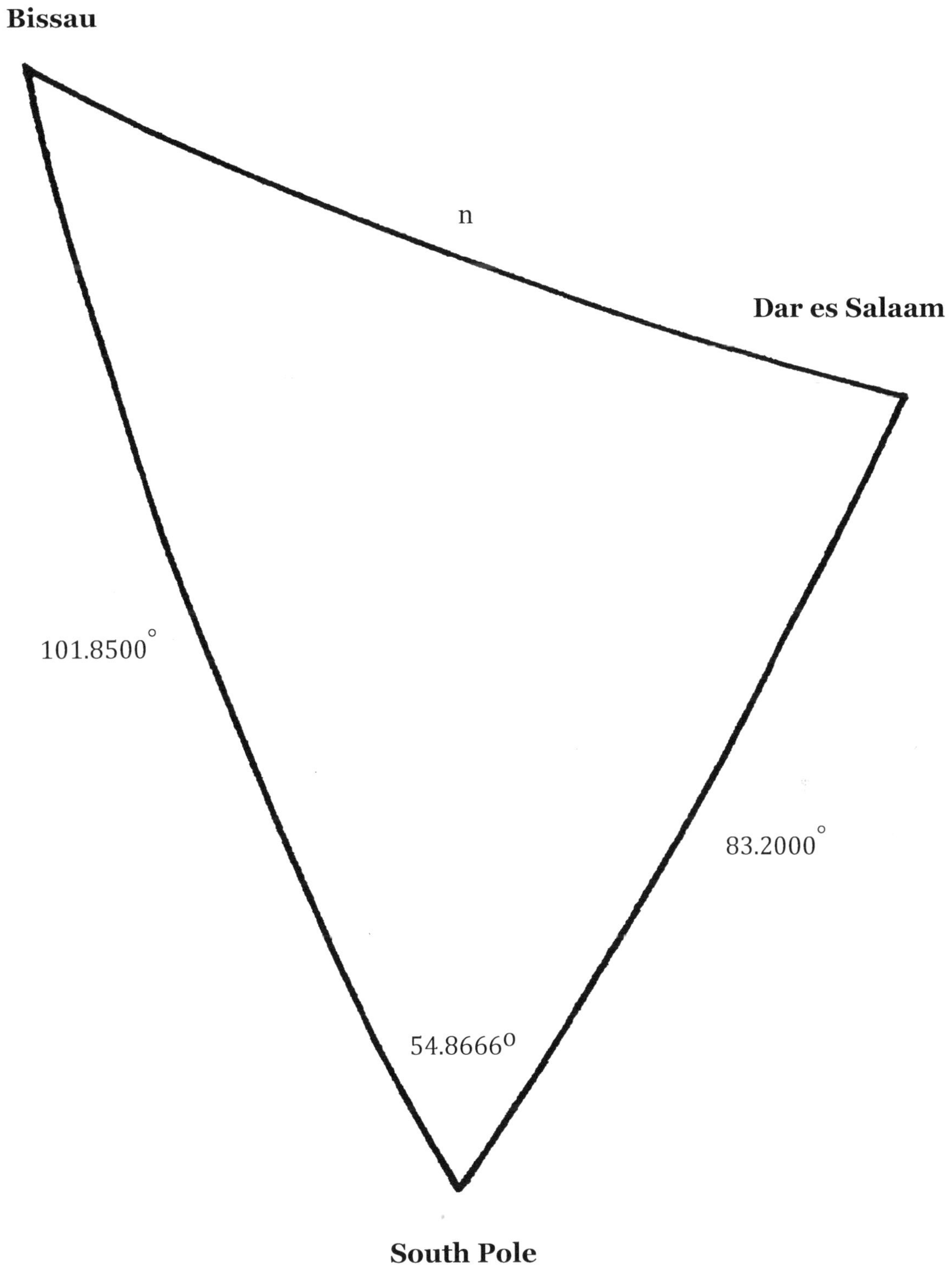
Bissau
n
Dar es Salaam
101.8500°
83.2000°
54.8666°
South Pole

Example 3

Mbabane, Swaziland is located at $26^o19'$ S $= \left(26 + {}^{19}/_{60}\right)^o$ S $\approx 26.3167^o$S, $31^o8'$ E $= \left(31 + {}^{8}/_{60}\right)^o$ E $\approx 31.1333^o$E.

Antananarivo, Madagascar is located at $18^o55'$ S $= \left(18 + {}^{55}/_{60}\right)^o$ S $\approx 18.9167^o$S, $47^o31'$ E $= \left(47 + {}^{31}/_{60}\right)^o$ E $\approx 47.5167^o$E.

Two sides of the spherical triangle are $90^o - 26.3167^o = 63.6833^o$ and $90^o - 18.9167^o = 71.0833^o$, respectively. The angle at the South Pole is $47.5167^o - 31.1333^o = 16.3834^o$.

See the illustration on the next page.

It follows that

$\cos n = \cos(63.6833^o)\cos(71.0833^o) + \sin(63.6833^o)\ \sin(71.0833^o)\cos(16.3834^o)$

$\cos n \approx\ 0.957240946572 \Rightarrow\ n \approx \cos^{-1}(0.957240946572)\ \Rightarrow\ n \approx 16.8155644284^o$.

Therefore, the approximate distance between the two capital cities is
$(16.8155644284^o)\left({}^{\pi}/_{180^o}\right)(3959 \text{ miles})$

$\approx (0.293486964856)(3959 \text{ miles})$

$\approx\ 1{,}162$ miles.

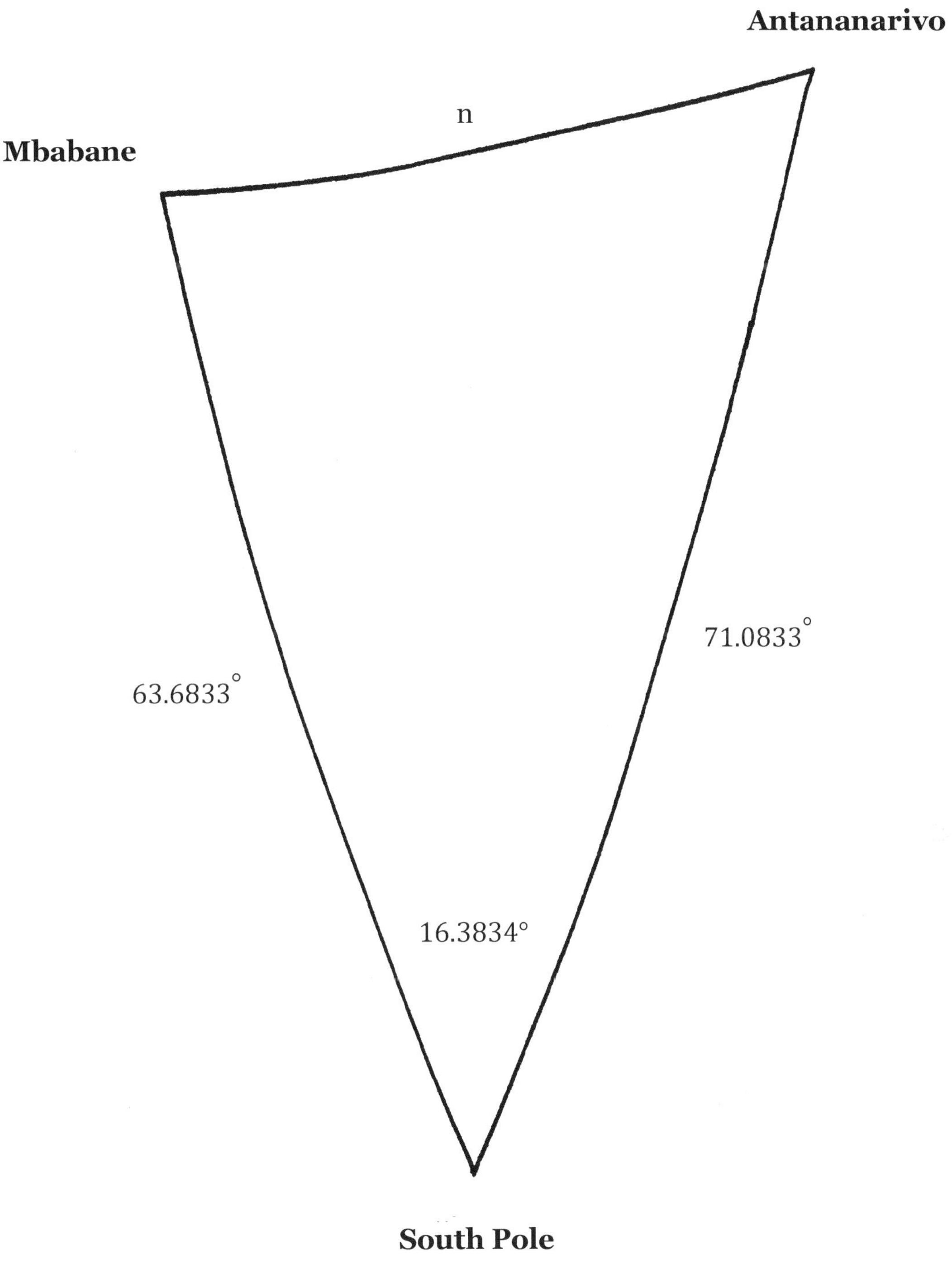
Antananarivo
n
Mbabane
71.0833°
63.6833°
16.3834°
South Pole

Chapter 2 Problems:

Approximate the distance between the following capitals in Africa.

1. Algiers, Algeria ($36^o45'$ N $3^o3'$ E)
   to Addis Ababa, Ethiopia ($9^o2'$ N $38^o42'$ E)

2. Luanda, Angola ($8^o50'$ S $13^o13'$ E)
   to Asmare, Eritrea ($15^o20'$ N $38^o56'$ E)

3. Porto-Novo, Benin ($6^o29'$ N $2^o37'$ E)
   to Libreville, Gabon ($0^o23'$ N $9^o27'$ E)

4. Gaborone, Botswana ($24^o38'$ S $25^o54'$ E)
   to Harare, Zimbabwe ($17^o49'$ S $31^o2'$ E)

5. Ouagadougou, Burkina Faso ($12^o22'$ N $1^o31'$ W)
   to Kampala, Uganda ($0^o19'$ N $32^o33'$ E)

6. Yaoundé, Cameroon ($3^o52'$ N $11^o31'$ E)
   to Lusaka, Zambia ($15^o25'$ S $28^o17'$ E)

7. Praia, Cabo Verde ($14^o55'$ N $23^o31'$ W)
   to N'Djamena, Chad ($12^o6'$ N $15^o2'$ E)

8. Yamoussoukro, Cote d'Ivoire ($6^o49'$ N $5^o16'$ W)
   to Tripoli, Libya ($32^o53'$ N $13^o10'$ E)

9. Kinshasa, Democratic Republic of the Congo ($4^o19'$ S $15^o18'$ E)
   to Nairobi, Kenya ($1^o17'$S $36^o49'$E)

10. Banjul, The Gambia (13°27′ N 16°34′ W)
to São Tomé, São Tomé and Principe (0°20′ N 6°44′ E)

11. Accra, Ghana (5°33′ N 0°13′ W)
to Moroni, Comoros(11°42′ S 43°14′ E)

12. Conakry, Guinea (9°30′ N 13°42′ W)
to Abuja, Nigeria (9°5′ N 7°32′ E)

13. Maseru, Lesotho (29°19′ S 27°29′ E)
to Lilongwe, Malawi (13°58′ S 33°47′ E)

14. Monrovia, Liberia (6°18′ N 10°48′ W)
to Khartoum, Sudan (15°36′ N 32°32′ E)

15. Bamako, Mali (12°39′ N 8°0′ W)
to Bangui, Central African Republic (4°22′ N 18°35′ E)

16. Nouakchott, Mauritania (18°4′ N 15°58′ W)
to Victoria, Seychelles (4°37′ S 55°27′ E)

17. Rabat, Morocco (34°1′ N 6°49′ W)
to Windhoek, Namibia (22°34′ S 17°5′ E)

18. Niamey, Niger (13°31′ N 2°7′ E)
to Kigali, Rwanda (1°57′ S 30°3′ E)

19. Brazzaville, Republic of the Congo (4°15′ S 15°17′ E)
to Djibouti, Djibouti (11°35′ N 43°9′ E)

20. Dakar, Senegal (14°44′ N 17°38′ W)
to Cairo, Egypt (30°3′ N 31°15′ E)

21. Freetown, Sierra Leone (8°29′ N 13°14′ W)
to Bujumbura, Burundi (3°22′ S 29°21′ E)

22. Juba, South Sudan (4°51′ N 31°37′ E)
to Mogadishu, Somalia (2°4′ N 45°20′ E)

23. Lomé, Togo (6°7′ N 1°13′ E)
to Maputo, Mozambique (25°57′ S 32°35′ E)

24. Tunis, Tunisia (36°48′ N 10°11′ E)
to Port Louis, Mauritius (20°9′ S 57°29′ E)

Chapter 2

Problem 1

Solution

Algiers, Algeria is located at $36^o45' \text{ N} = \left(36 + {}^{45}/_{60}\right)^o \text{ N} = 36.7500^o \text{ N}$,
$3^o3' \text{ E} = \left(3 + {}^{3}/_{60}\right)^o \text{ E} = 3.0500^o \text{ E}$.

Addis Ababa, Ethiopia is located at $9^o2' \text{ N} = \left(9 + {}^{2}/_{60}\right)^o \text{ N} \approx 9.0333^o \text{ N}$,
$38^o42' \text{ E} = \left(38 + {}^{42}/_{60}\right)^o \text{ E} = 38.7000^o \text{ E}$.

Two sides of the spherical triangle are $90^o - 36.7500^o = 53.2500^o$ and $90^o - 9.0333^o = 80.9667^o$, respectively. The angle at the North Pole is $38.7000^o - 3.0500^o = 35.6500^o$.

See the illustration on the next page.

It follows that

$$\cos n = \cos(53.2500^o)\cos(80.9667^o) + \sin(53.2500^o)\sin(80.9667^o)\cos(35.6500^o)$$

$$\cos n \approx 0.736959498133 \Rightarrow n \approx \cos^{-1}(0.736959498133) \Rightarrow n \approx 42.5269485004^o.$$

Therefore, the approximate distance between the two capital cities is

$$(42.5269485004^o)\left({}^{\pi}/_{180^o}\right)(3959 \text{ miles})$$

$$\approx (0.742235272158)(3959 \text{ miles})$$

$$\approx 2{,}939 \text{ miles.}$$

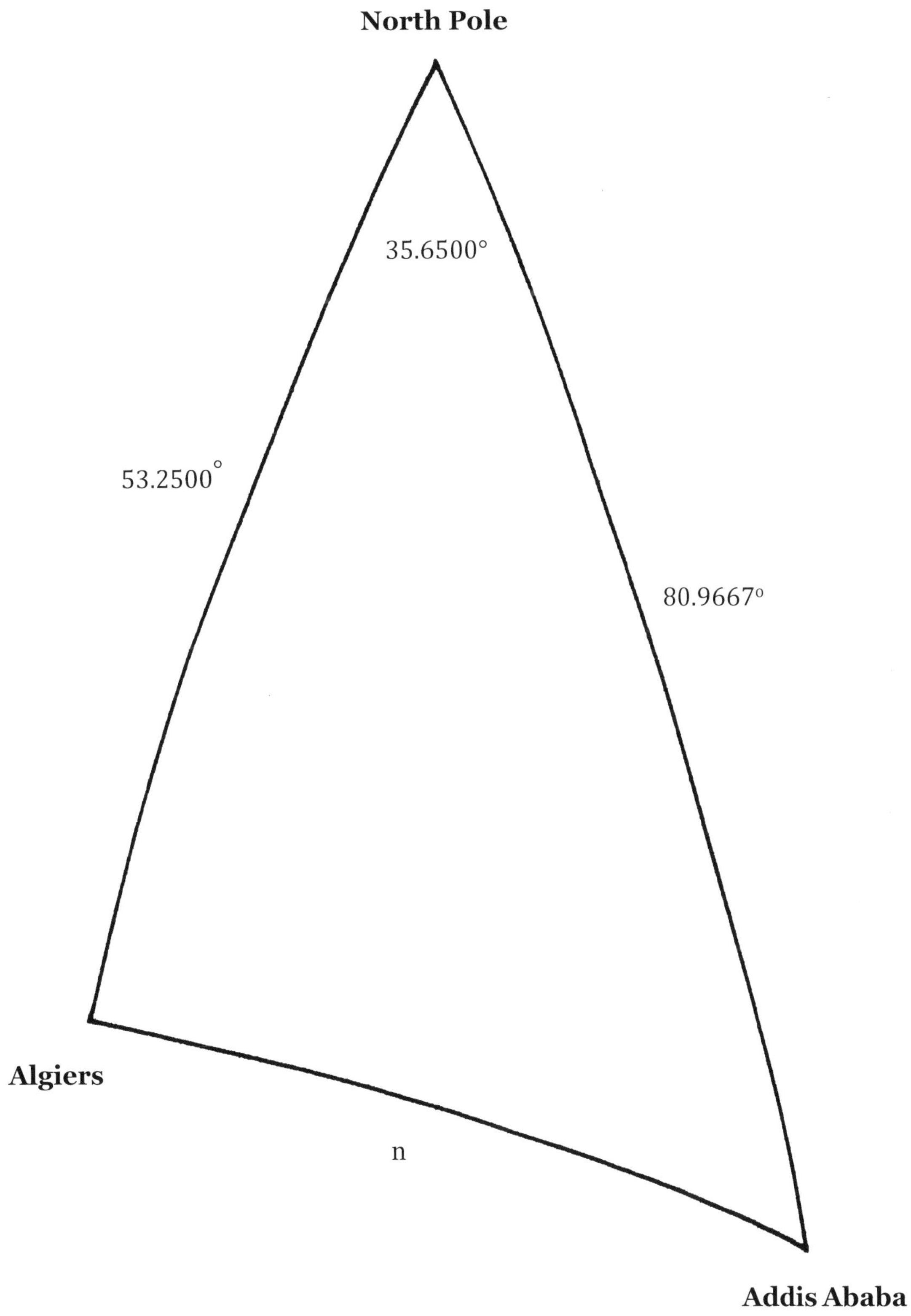
North Pole
35.6500°
53.2500°
80.9667°
Algiers
n
Addis Ababa

Chapter 2

Problem 2

Solution

Luanda, Angola is located at $8^o50'\ \mathrm{S} = \left(8 + {}^{50}/_{60}\right)^o \mathrm{S} \approx 8.8333^o\ \mathrm{S}$,

$13^o13'\ \mathrm{E} = \left(13 + {}^{13}/_{60}\right)^o\ \mathrm{E} \approx 13.2167^o\ \mathrm{E}$.

Asmara, Eritrea is located at $15^o20'\ \mathrm{N} = \left(15 + {}^{20}/_{60}\right)^o\ \mathrm{N} \approx 15.3333^o\ \mathrm{N}$, $38^o56'$

$\mathrm{E} = \left(38 + {}^{56}/_{60}\right)^o\ \mathrm{E} \approx 38.9333^o\ \mathrm{E}$.

Two sides of the spherical triangle are $90^o + 8.8333^o = 98.8333^o$ and $90^o - 15.3333^o = 74.6667^o$, respectively. The angle at the North Pole is $38.9333^o - 13.2167^o = 25.7166^o$.

See the illustration on the next page.

It follows that

$$\cos n = \cos(98.8333^o)\cos(74.6667^o) + \sin(98.8333^o)\ \sin(74.6667^o)\cos(25.7166^o)$$

$$\cos n \approx 0.817968978619 \Rightarrow n \approx \cos^{-1}(0.817968978619) \Rightarrow n \approx 35.1180054986^o.$$

Therefore, the approximate distance between the two capital cities is

$$(35.1180054986^o)\left({}^{\pi}/_{180^o}\right)(3959 \text{ miles})$$

$$\approx (0.612924822685)(3959 \text{ miles})$$

$$\approx 2{,}427 \text{ miles.}$$

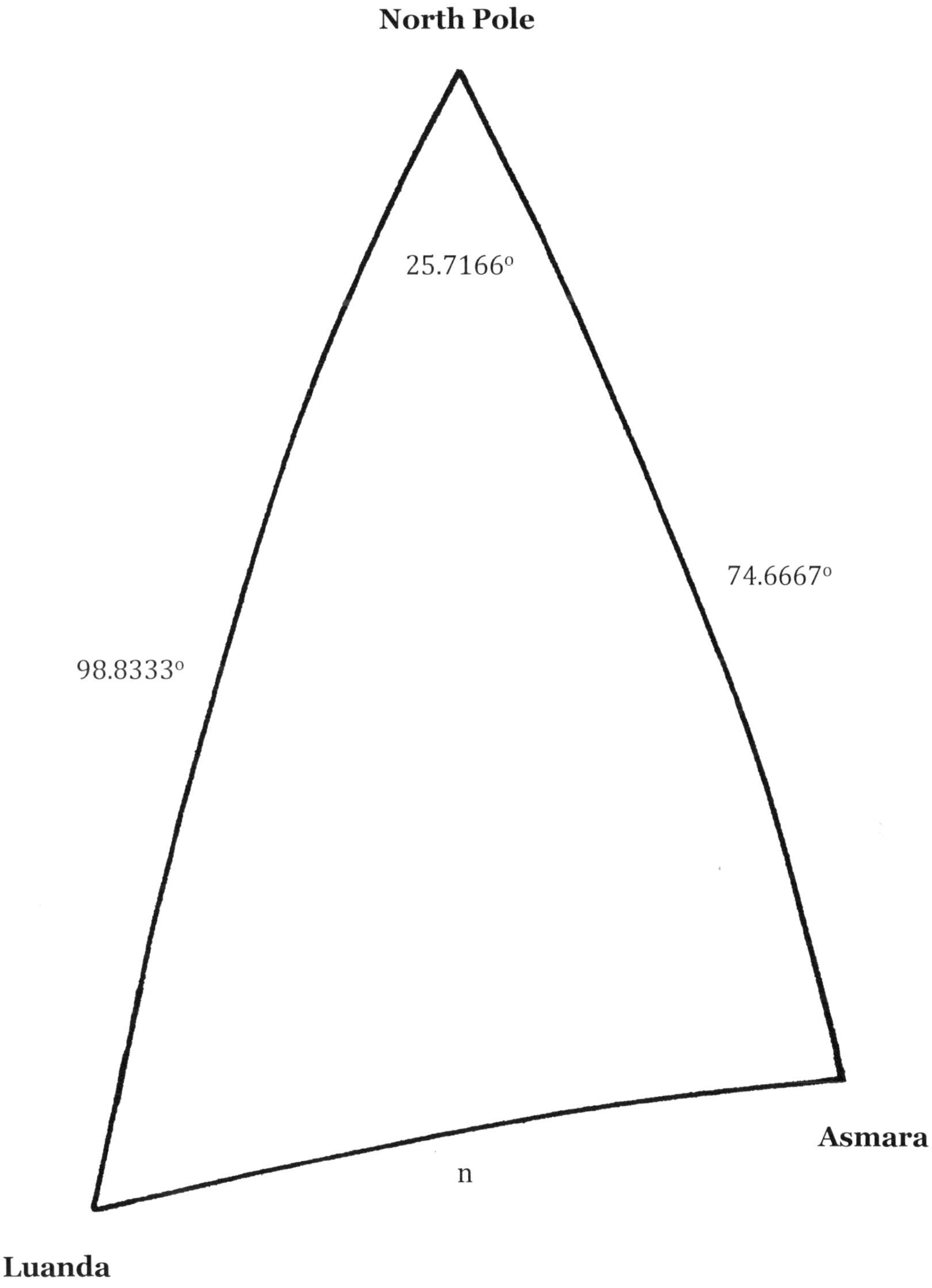
North Pole
25.7166°
74.6667°
98.8333°
Asmara
n
Luanda

Chapter 2

Problem 3

Solution

Porto-Novo, Benin is located at $6^{o}29' \text{ N} = \left(6 + {}^{29}/_{60}\right)^{o} \text{ N} \approx 6.4833^{o} \text{ N}$,

$2^{o}37' \text{ E} = \left(2 + {}^{37}/_{60}\right)^{o} \text{ E} \approx 2.6167^{o} \text{ E}$.

Libreville, Gabon is located at $0^{o}23' \text{ N} = \left({}^{23}/_{60}\right)^{o} \text{ N} \approx 0.3833^{o} \text{ N}$,

$9^{o}27' \text{ E} = \left(9 + {}^{27}/_{60}\right)^{o} \text{ E} = 9.4500^{o} \text{ E}$.

Two sides of the spherical triangle are $90^{o} - 6.4833^{o} = 83.5167^{o}$ and $90^{o} - 0.3833^{o} = 89.6167^{o}$, respectively. The angle at the North Pole is $9.4500^{o} - 2.6167^{o} = 6.8333^{o}$.

See the illustration on the next page.

It follows that

$\cos \text{n} = \cos(83.5167^{o}) \cos(89.6167^{o}) + \sin(83.5167^{o}) \sin(89.6167^{o}) \cos(6.8333^{o})$

$\cos \text{n} \approx 0.987280054807 \Rightarrow \text{n} \approx \cos^{-1}(0.987280054807) \Rightarrow \text{n} \approx 9.14832695812^{o}$.

Therefore, the approximate distance between the two capital cities is
$(9.14832695812^{o})\left({}^{\pi}/_{180^{o}}\right)(3959 \text{ miles})$

$\approx (0.159668426468)(3959 \text{ miles})$

≈ 632 miles.

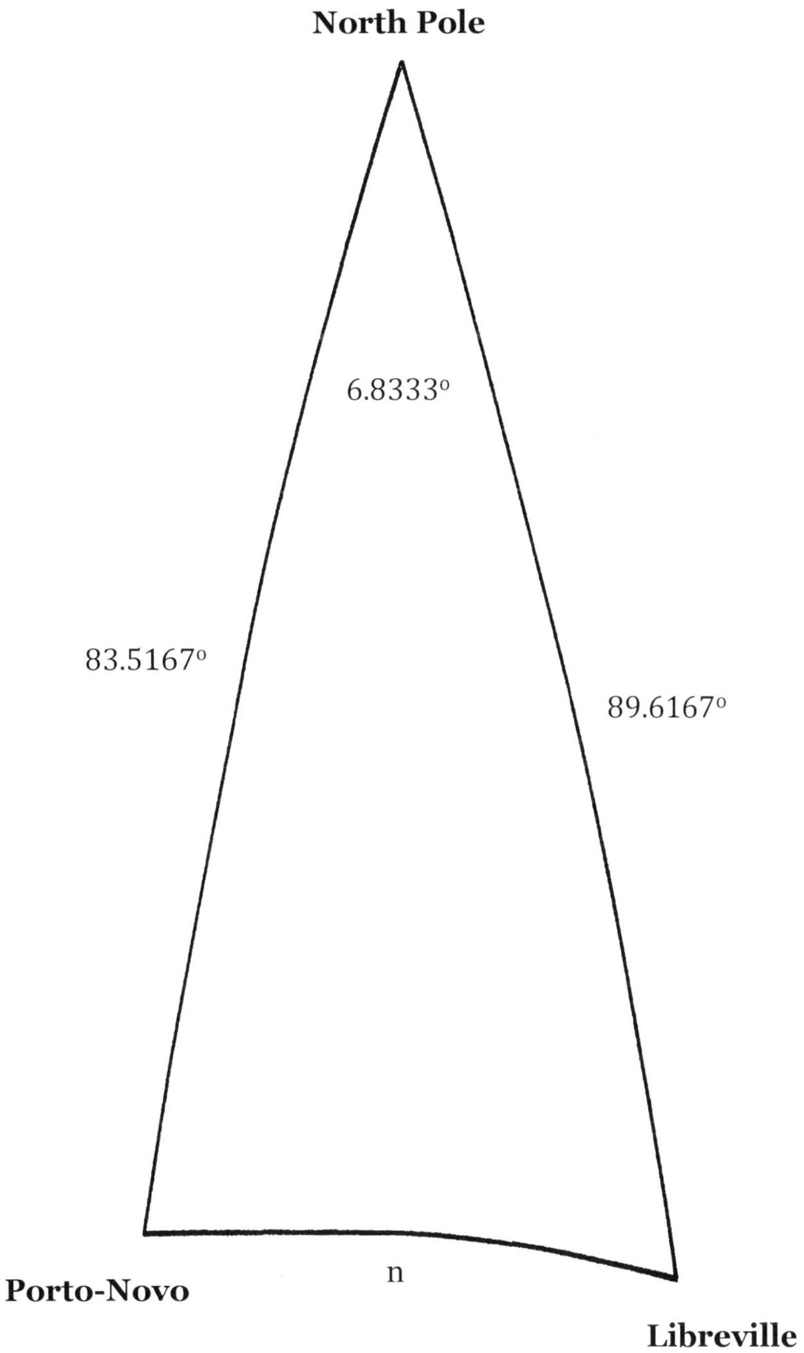
North Pole
6.8333°
83.5167°
89.6167°
n
Porto-Novo
Libreville

Chapter 2
Problem 4
Solution

Gaborone, Botswana is located at $24^o38'\ S = \left(24 + {}^{38}/_{60}\right)^o\ S \approx 24.6333^o\ S$, $25^o54'\ E = \left(25 + {}^{54}/_{60}\right)^o\ E = 25.9000^o\ E$.

Harare, Zimbabwe is located at $17^o49'\ S = \left(17 + {}^{49}/_{60}\right)^o\ S \approx 17.8167^o\ S$, $31^o2'\ E = \left(31 + {}^{2}/_{60}\right)^o\ E \approx 31.0333^o\ E$.

Two sides of the spherical triangle are $90^o - 24.6333^o = 65.3667^o$ and $90^o - 17.8167^o \approx 72.1833^o$, respectively. The angle at the South Pole is $31.0333^o - 25.9000^o = 5.1333^o$.

See the illustration on the next page.

It follows that

$$\cos n = \cos(65.3667^o)\cos(72.1833^o) + \sin(65.3667^o)\ \sin(72.1833^o)\cos(5.1333^o)$$

$$\cos n \approx 0.989460247964 \Rightarrow n \approx \cos^{-1}(0.989460247964) \Rightarrow n \approx 8.32597327559^o.$$

Therefore, the approximate distance between the two capital cities is

$$(8.32597327559^o)\left({}^{\pi}/_{180^o}\right)(3959\ \text{miles})$$

$$\approx (0.145315647092)(3959\ \text{miles})$$

$$\approx 575\ \text{miles}.$$

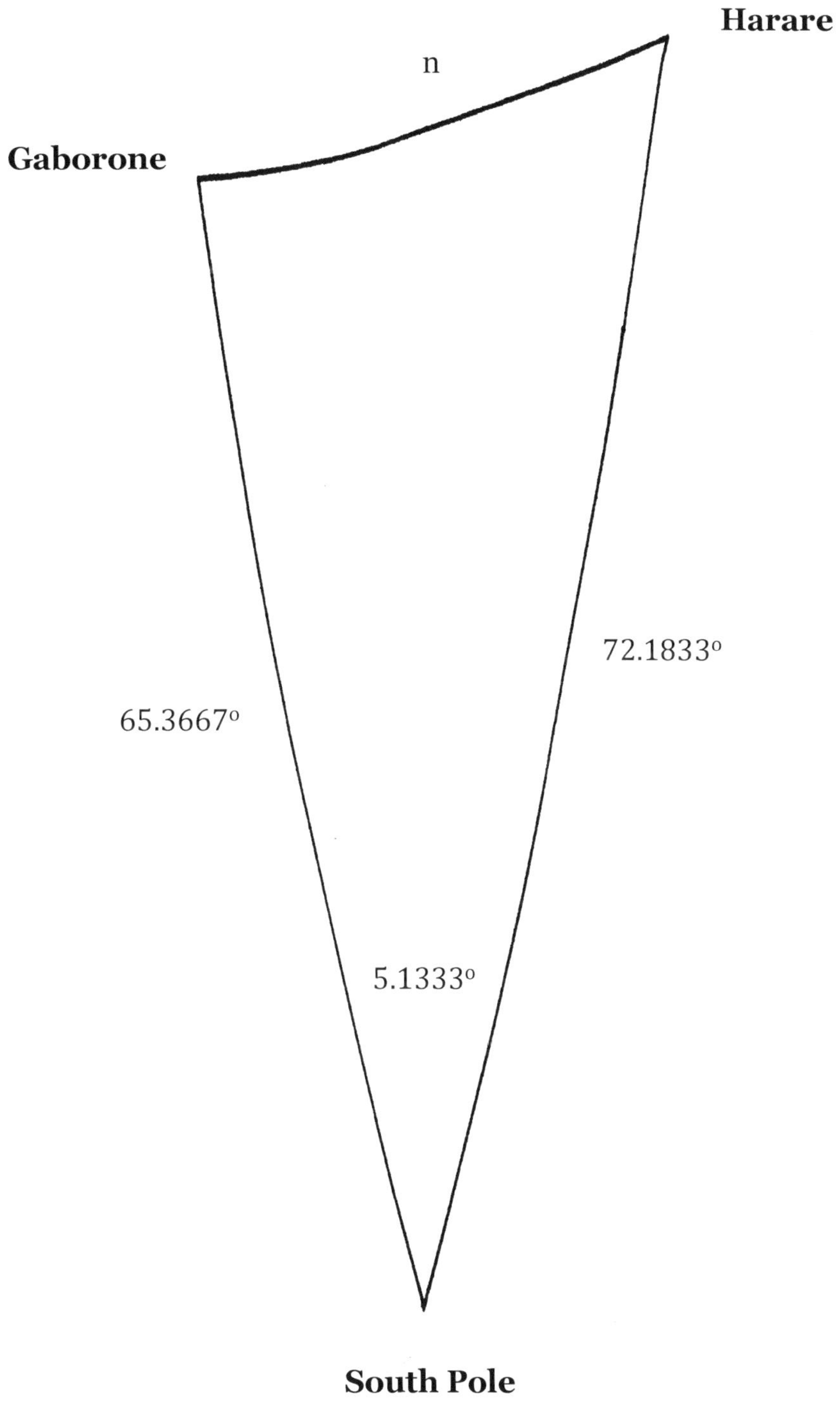
Harare
n
Gaboronen
72.1833°
65.3667°
5.1333°
South Pole

Chapter 2
Problem 5
Solution

Ouagadougou, Burkina Faso is located at

$12^o22'\ \mathrm{N} = \left(12 + {}^{22}/_{60}\right)^o\ \mathrm{N} \approx 12.3667^o\ \mathrm{N}$, $1^o31'\ \mathrm{W} = \left(1 + {}^{31}/_{60}\right)^o\ \mathrm{W} \approx 1.5167^o\ \mathrm{W}$.

Kampala, Uganda is located at

$0^o19'\ \mathrm{N} = \left({}^{19}/_{60}\right)^o\ \mathrm{N} \approx 0.3167^o\ \mathrm{N}$,$32^o33'\ \mathrm{E} = \left(32 + {}^{33}/_{60}\right)^o\ \mathrm{E} \approx 32.5500^o\ \mathrm{E}$.

Two sides of the spherical triangle are $90^o - 12.3667^o = 77.6333^o$ and $90^o - 0.3167^o = 89.6833^o$, respectively. The angle at the North Pole is $1.5167^o + 32.5500^o = 34.0667^o$.

See the illustration on the next page.

It follows that

$\cos n = \cos(77.6333^o)\cos(89.6833^o) + \sin(77.6333^o)\ \sin(89.6833^o)\cos(34.0667^o)$

$\cos n \approx 0.810336360238 \Rightarrow n \approx \cos^{-1}(0.810336360238) \Rightarrow n \approx 35.8711922515^o$.

Therefore, the approximate distance between the two capital cities is
$(35.8711922515^o)\left({}^{\pi}/_{180^o}\right)(3959 \text{ miles})$

$\approx (0.626070411405)(3959 \text{ miles})$

$\approx 2{,}479$ miles.

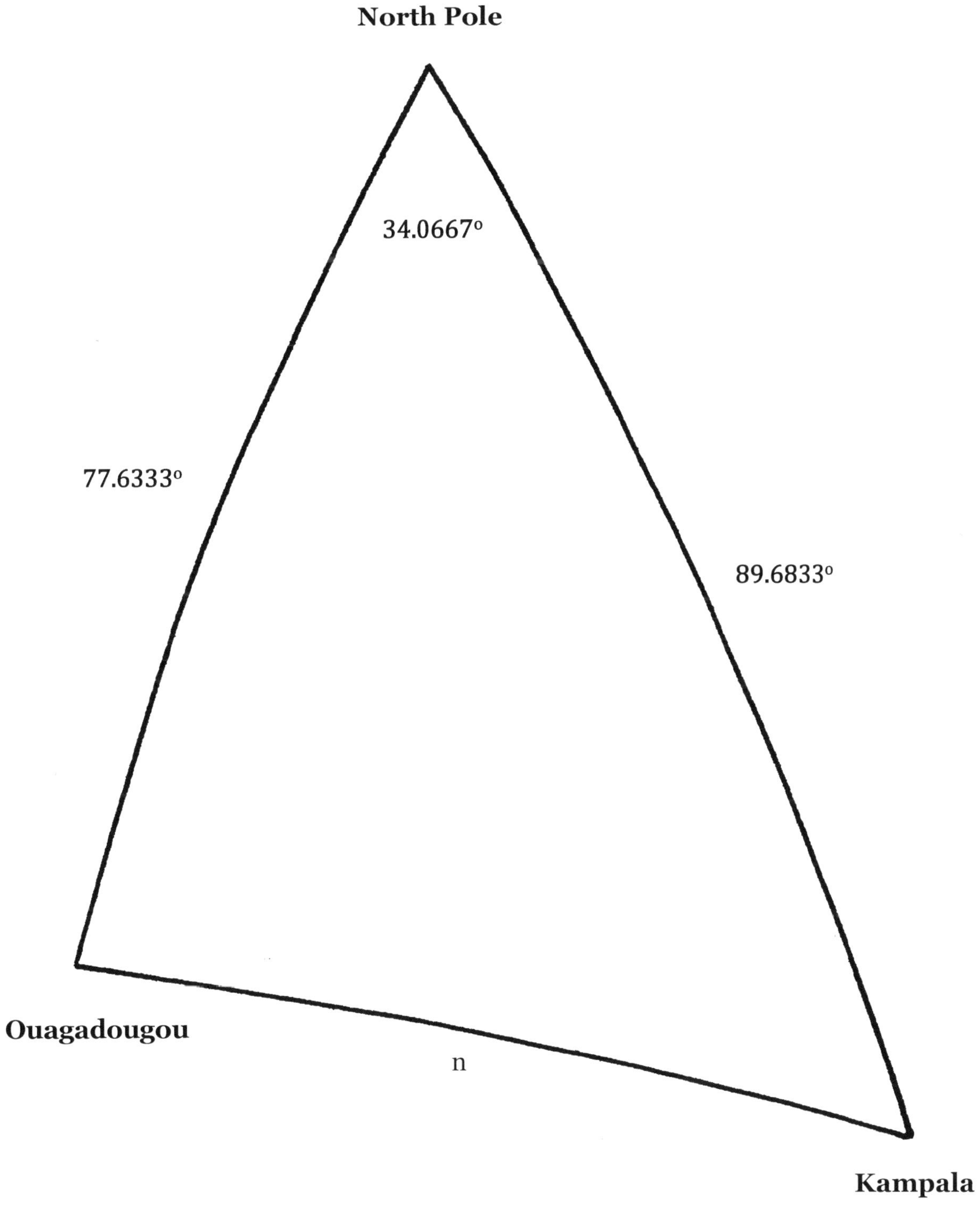
North Pole
34.0667°
77.6333°
89.6833°
Ouagadougou
n
Kampala

Chapter 2
Problem 6
Solution

Yaoundé, Cameroon is located at $3^o52'\ \mathrm{N} = \left(3 + {}^{52}/_{60}\right)^o\ \mathrm{N} \approx 3.8667^o\ \mathrm{N}$,

$11^o31'\ \mathrm{E} = \left(11 + {}^{31}/_{60}\right)^o\ \mathrm{E} \approx 11.5167^o\ \mathrm{E}$.

Lusaka, Zambia is located at $15^o25'\ \mathrm{S} = \left(15 + {}^{25}/_{60}\right)^o\ \mathrm{S} \approx 15.4167^o\ \mathrm{S}$,

$28^o17'\ \mathrm{E} = \left(28 + {}^{17}/_{60}\right)^o\ \mathrm{E} \approx 28.2833^o\ \mathrm{E}$.

Two sides of the spherical triangle are $90^o + 3.8667^o = 93.8667^o$ and $90^o - 15.4167^o = 74.5833^o$, respectively. The angle at the South Pole is $28.2833^o - 11.5167^o \approx 16.7666^o$.

See the illustration on the next page.

It follows that

$\cos n = \cos(93.8667^o)\cos(74.5833^o) + \sin(93.8667^o)\ \sin(74.5833^o)\cos(16.7666^o)$

$\cos n \approx 0.903007458883 \Rightarrow n \approx \cos^{-1}(0.903007458883) \Rightarrow n \approx 25.4437558295^o$.

Therefore, the approximate distance between the two capital cities is
$(25.4437558295^o)\left({}^{\pi}/_{180^o}\right)(3959\ \text{miles})$

$\approx (0.444077313298)(3959\ \text{miles})$

$\approx\ 1{,}758\ \text{miles}$.

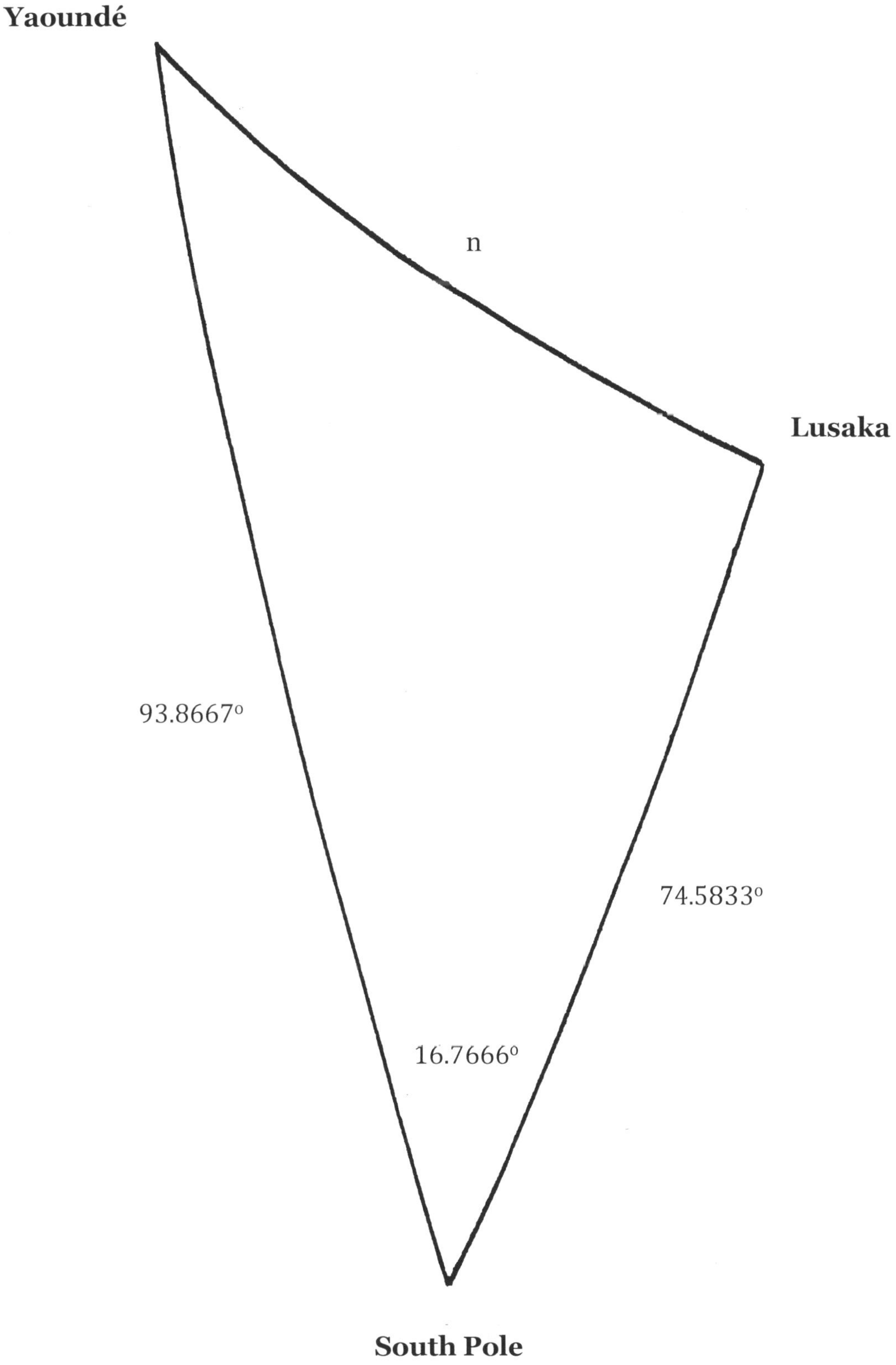
Yaoundé
n
Lusaka
93.8667°
74.5833°
16.7666°
South Pole

Chapter 2
Problem 7
Solution

Praia, Cabo Verde is located at $14^o55' \text{ N} = \left(14 + {}^{55}/_{60}\right)^o \text{ N} \approx 14.9167^o \text{ N}$,

$23^o31' \text{ W} = \left(23 + {}^{31}/_{60}\right)^o \text{ W} \approx 23.5167^o \text{ W}$.

N'Djamena, Chad is located at $12^o6' \text{ N} = \left(12 + {}^{6}/_{60}\right)^o \text{ N} = 12.1000^o \text{ N}$,

$15^o2' \text{ E} = \left(15 + {}^{2}/_{60}\right)^o \text{ E} \approx 15.0333^o \text{ E}$.

Two sides of the spherical triangle are $90^o - 14.9167^o = 75.0833^o$ and $90^o - 12.1000^o = 77.9000^o$, respectively. The angle at the North Pole is $23.5167^o + 15.0333^o = 38.5500^o$.

See the illustration on the next page.

It follows that

$\cos n = \cos(75.0833^o)\cos(77.9000^o) + \sin(75.0833^o)\ \sin(77.9000^o)\cos(38.5500^o)$

$\cos n \approx 0.792879308617 \ \Rightarrow\ n \approx \cos^{-1}(0.792879308617) \ \Rightarrow\ n \approx 37.544592592^o$.

Therefore, the approximate distance between the two capital cities is $(37.544592592^o)\left({}^{\pi}/_{180^o}\right)(3959 \text{ miles})$

$\approx (0.65527675705)(3959 \text{ miles})$

$\approx\ 2{,}594\ \text{ miles.}$

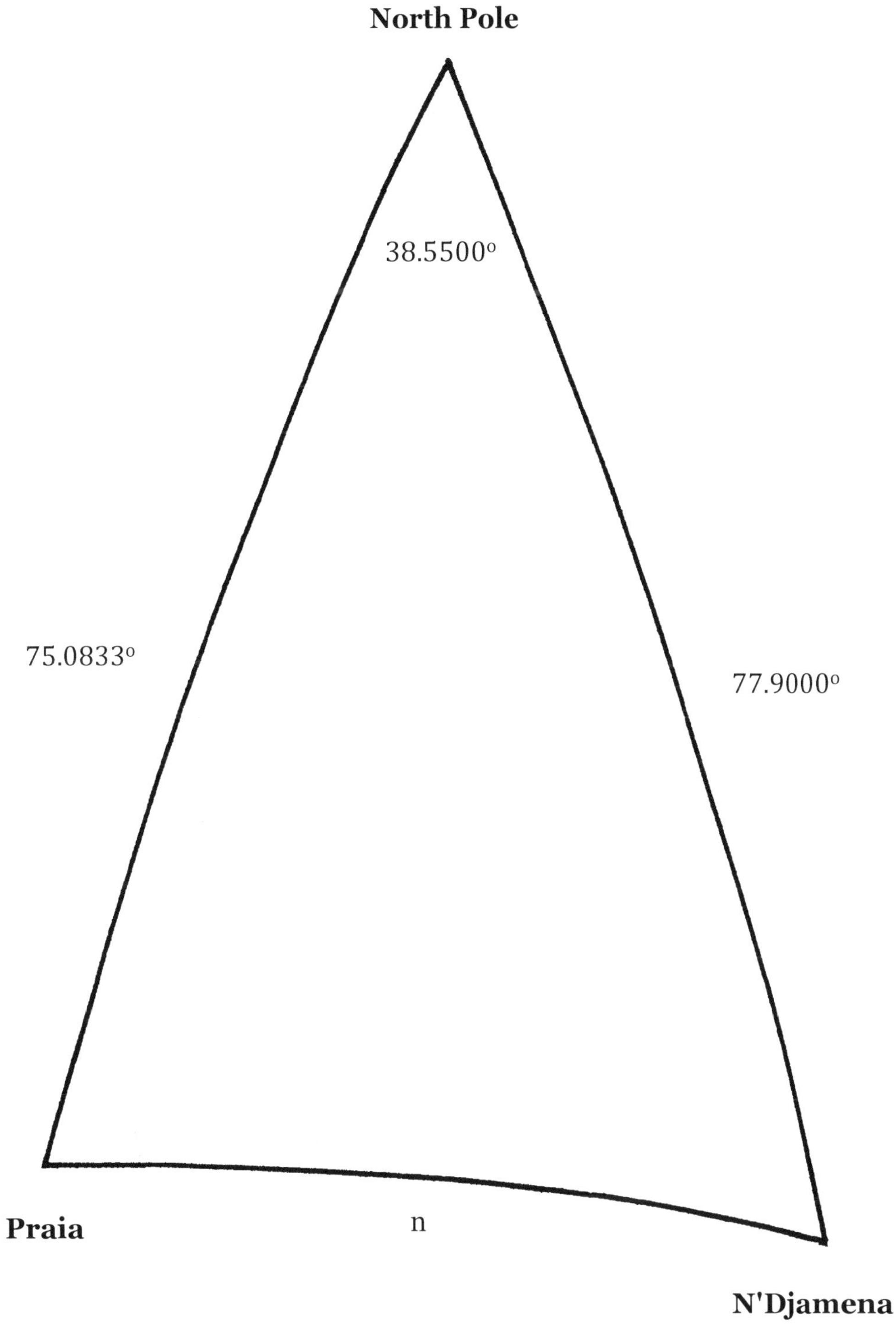
North Pole
38.5500°
75.0833°
77.9000°
Praia
n
N'Djamena

Chapter 2
Problem 8
Solution

Yamoussoukro, Cote d'Ivoire is located at $6^o 49' \text{ N} = \left(6 + {}^{49}/_{60}\right)^o \text{ N} \approx 6.8167^o \text{ N}$, $5^o 16' \text{ W} = \left(5 + {}^{16}/_{60}\right)^o \text{ W} \approx 5.2667^o \text{ W}$.

Tripoli, Libya is located at $32^o 53' \text{ N} = \left(32 + {}^{53}/_{60}\right)^o \text{ N} \approx 32.8833^o \text{ N}$, $13^o 10' \text{ E} = \left(13 + {}^{10}/_{60}\right)^o \text{ E} \approx 13.1667^o \text{ E}$.

Two sides of the spherical triangle are $90^o - 6.8167^o = 83.1833^o$ and $90^o - 32.8833^o = 57.1167^o$, respectively. The angle at the North Pole is $5.2667^o + 13.1667^o = 18.4334^o$.

See the illustration on the next page.

It follows that

$\cos n = \cos(83.1833^o)\cos(57.1167^o) + \sin(83.1833)\ \sin(57.1167^o)\cos(18.4334^o)$

$\cos n \approx 0.855501001995 \Rightarrow n \approx \cos^{-1}(0.855501001995) \Rightarrow n \approx 31.1848725296^o$.

Therefore, the approximate distance between the two capital cities is

$(31.1848725296^o)\left({}^{\pi}/_{180^o}\right)(3959 \text{ miles})$

$\approx (0.544278702456)(3959 \text{ miles})$

$\approx 2{,}155$ miles.

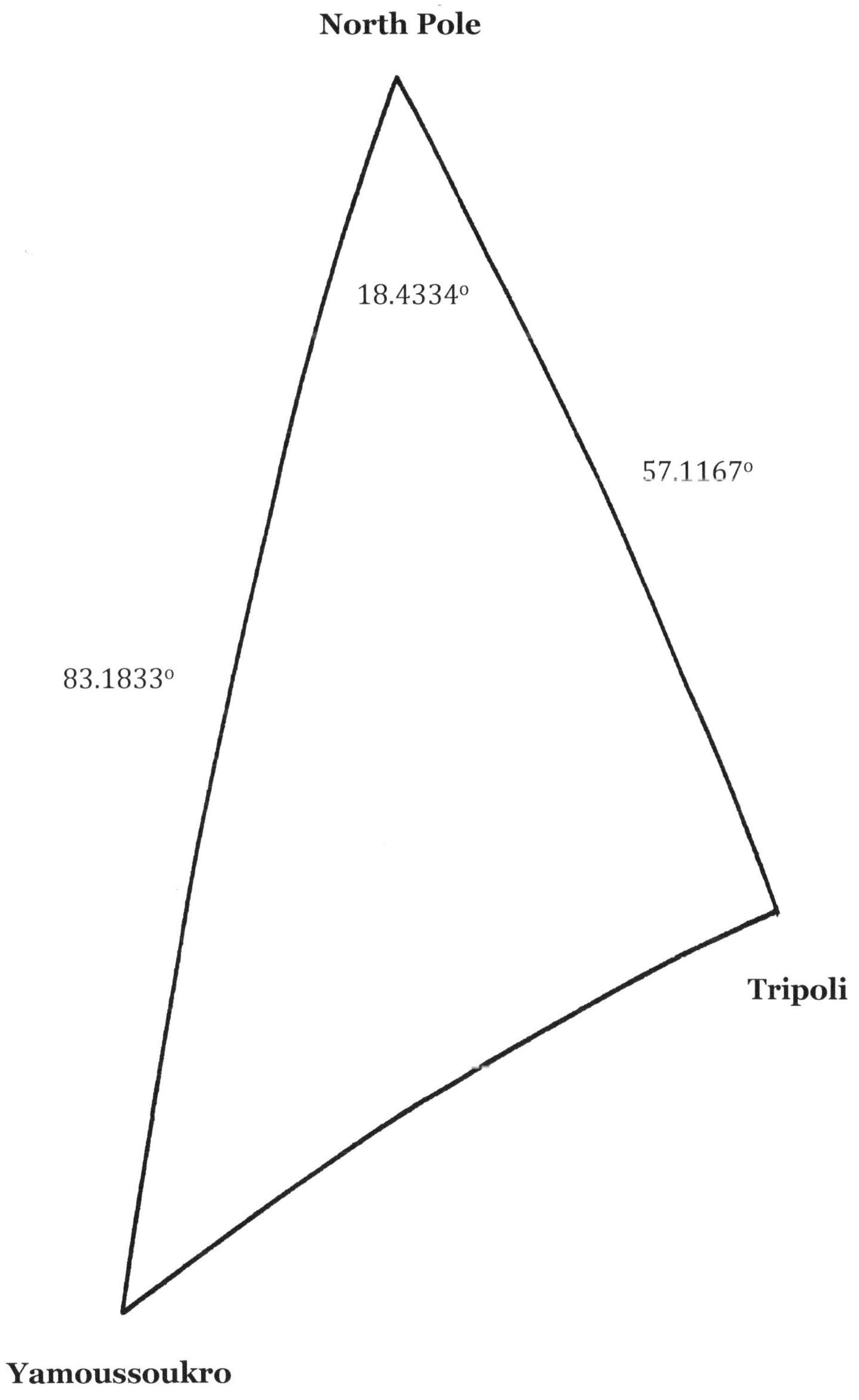
North Pole
18.4334°
57.1167°
83.1833°
Tripoli
Yamoussoukro

Chapter 2

Problem 9

Solution

Kinshasa, Democratic Republic of the Congo is located at

$4^{o}19'\ \mathrm{S} = \left(4 + {}^{19}/_{60}\right)^{o}\ \mathrm{S} \approx 4.3167^{o}\ \mathrm{S}$, $15^{o}18'\ \mathrm{E} = \left(15 + {}^{18}/_{60}\right)^{o}\ \mathrm{E} = 15.3000^{o}\ \mathrm{E}$.

Nairobi, Kenya is located at $1^{o}17'\ \mathrm{S} = \left(1 + {}^{17}/_{60}\right)^{o}\ \mathrm{S} \approx 1.2833^{o}\ \mathrm{S}$,

$36^{o}49'\ \mathrm{E} = \left(36 + {}^{49}/_{60}\right)^{o}\ \mathrm{E} \approx 36.8167^{o}\ \mathrm{E}$.

The sides of the spherical triangle are $90^{o} - 4.3167^{o} = 85.6833^{o}$ and $90^{o} - 1.2833^{o} = 88.7167^{o}$, respectively. The angle at the South Pole is $36.8167^{o} - 15.3000^{o} = 21.5167^{o}$.

See the illustration on the next page.

It follows that

$$\cos n = \cos(85.6833^{o})\cos(88.7167^{o}) + \sin(85.6833^{o})\ \sin(88.7167^{o})\cos(21.5167^{o})$$

$$\cos n \approx 0.929124682637 \ \Rightarrow\ n \approx \cos^{-1}(0.929124682637) \ \Rightarrow\ n \approx 21.7012224269^{o}.$$

Therefore, the approximate distance between the two capital cities is

$(21.7012224269^{o})\left({}^{\pi}/_{180^{o}}\right)(3959\ \text{miles})$

$\approx (0.378757783057)(3959\ \text{miles})$

$\approx\ 1{,}500\ \text{miles}.$

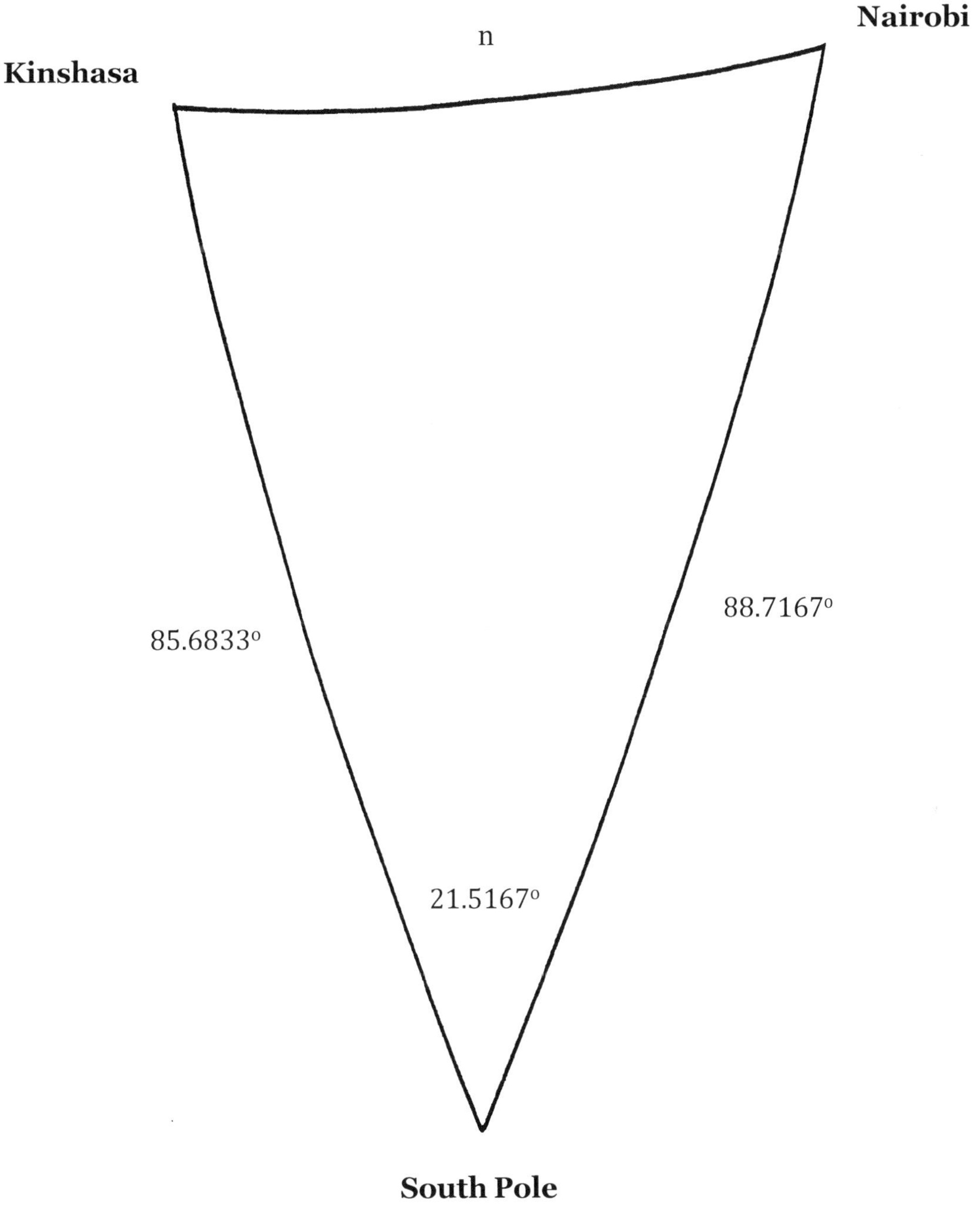
Kinshasa
n
Nairobi
85.6833°
88.7167°
21.5167°
South Pole

Chapter 2
Problem 10
Solution

Banjul, The Gambia is located at $13^o 27'$ N $= \left(13 + {}^{27}/_{60}\right)^o$ N $= 13.4500^o$ N ,

$16^o 34'$ W $= \left(16 + {}^{34}/_{60}\right)^o$ W $\approx 16.5667^o$ W .

São Tomé, São Tomé and Principe is located at $0^o 20'$ N $= \left({}^{20}/_{60}\right)^o$ N $\approx 0.3333^o$ N ,

$6^o 44'$ E $= \left(6 + {}^{44}/_{60}\right)^o$ E $\approx 6.7333^o$ E .

Two sides of the spherical triangle are $90^o - 13.4500^o = 76.5500^o$ and $90^o - 0.3333^o = 89.6667^o$, respectively. The angle at the North Pole is $16.5667^o + 6.7333^o = 23.3000^o$.

See the illustration on the next page.

It follows that

$\cos n = \cos(76.5500^o)\cos(89.6667^o) + \sin(76.5500^o)\ \sin(89.6667^o)\cos(23.3000^o)$

$\cos n \approx 0.89459433633 \ \Rightarrow\ n \approx \cos^{-1}(0.89459433633\ ) \ \Rightarrow\ n \approx 26.5436289658^o.$

Therefore, the approximate distance between the two capital cities is $(26.5436289658^o)\left({}^{\pi}/_{180^o}\right)(3959 \text{ miles})$

$\approx (0.46327372088)(3959 \text{ miles})$

$\approx$ 1,834 miles.

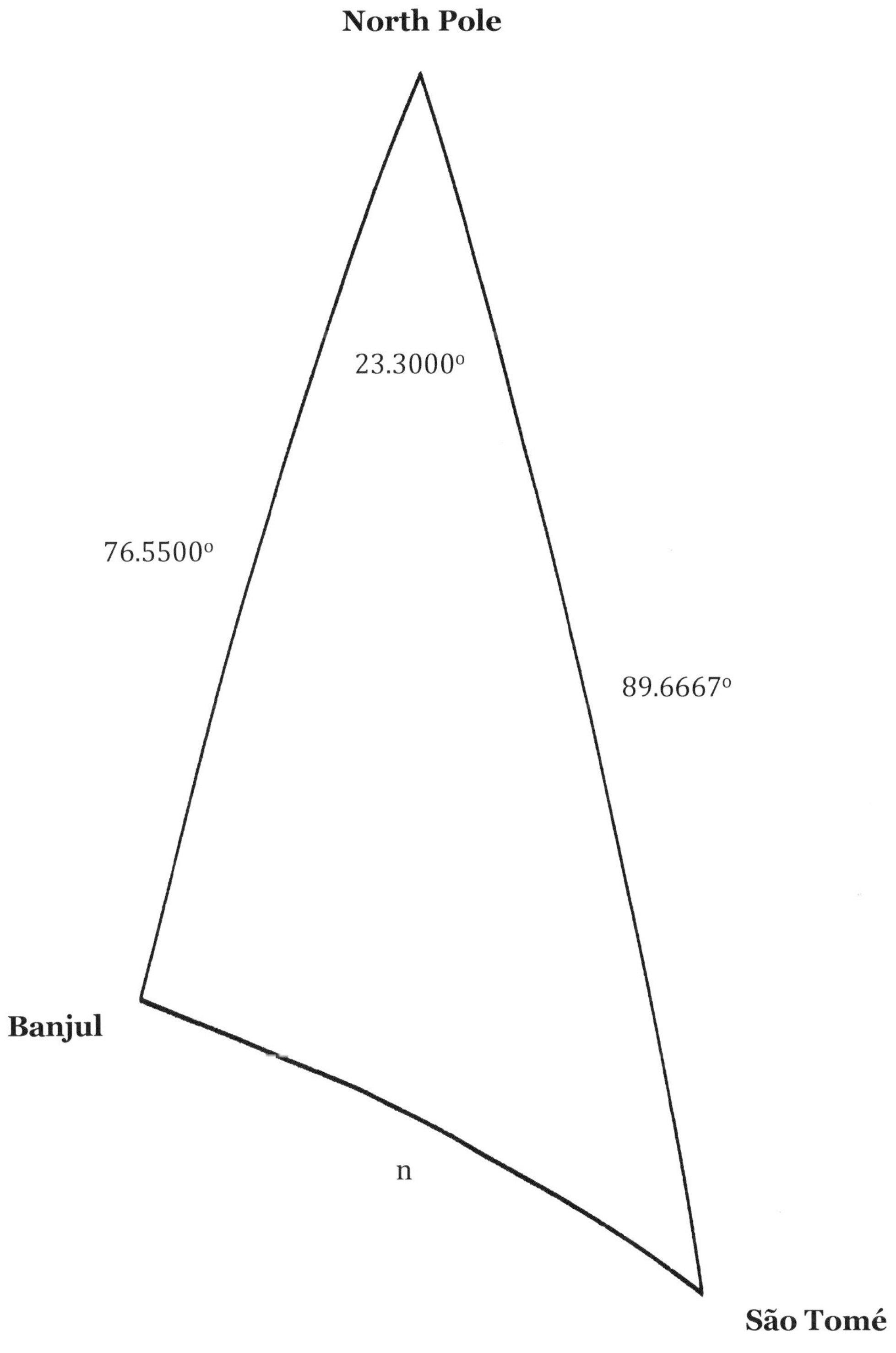
North Pole
23.3000°
76.5500°
89.6667°
Banjul
n
São Tomé

Chapter 2

Problem 11

Solution

Accra, Ghana is located at $5^o33'$ N $= \left(5 + {}^{33}/_{60}\right)^o$ N $= 5.5500^o$ N ,

$0^o13'$ W $= \left({}^{13}/_{60}\right)^o$ W $\approx 0.2167^o$ W .

Moroni, Comoros is located at $11^o42'$ S $= \left(11 + {}^{42}/_{60}\right)^o$ S $= 11.7000^o$ S ,

$43^o14'$ E $= \left(43 + {}^{14}/_{60}\right)^o$ E $\approx 43.2333^o$ E .

Two sides of the spherical triangle are $90^o - 5.5500^o = 84.4500^o$ and $90^o + 11.7000^o = 101.7000^o$, respectively. The angle at the North Pole is $0.2167^o + 43.2333^o = 43.4500^o$.

See the illustration on the next page.

It follows that

$\cos n = \cos(84.4500^o)\cos(101.7000^o) + \sin(84.4500^o)\ \sin(101.7000^o)\cos(43.4500^o)$

$\cos n \approx\ 0.687946106747\ \Rightarrow\ n \approx \cos^{-1}(0.687946106747)\ \Rightarrow\ n \approx 46.5322554795^o.$

Therefore, the approximate distance between the two capital cities is

$(46.5322554795^o)\left({}^{\pi}/_{180^o}\right)(3959\text{ miles})$

$\approx (0.812141066496)(3959\text{ miles})$

$\approx\ 3{,}215$ miles.

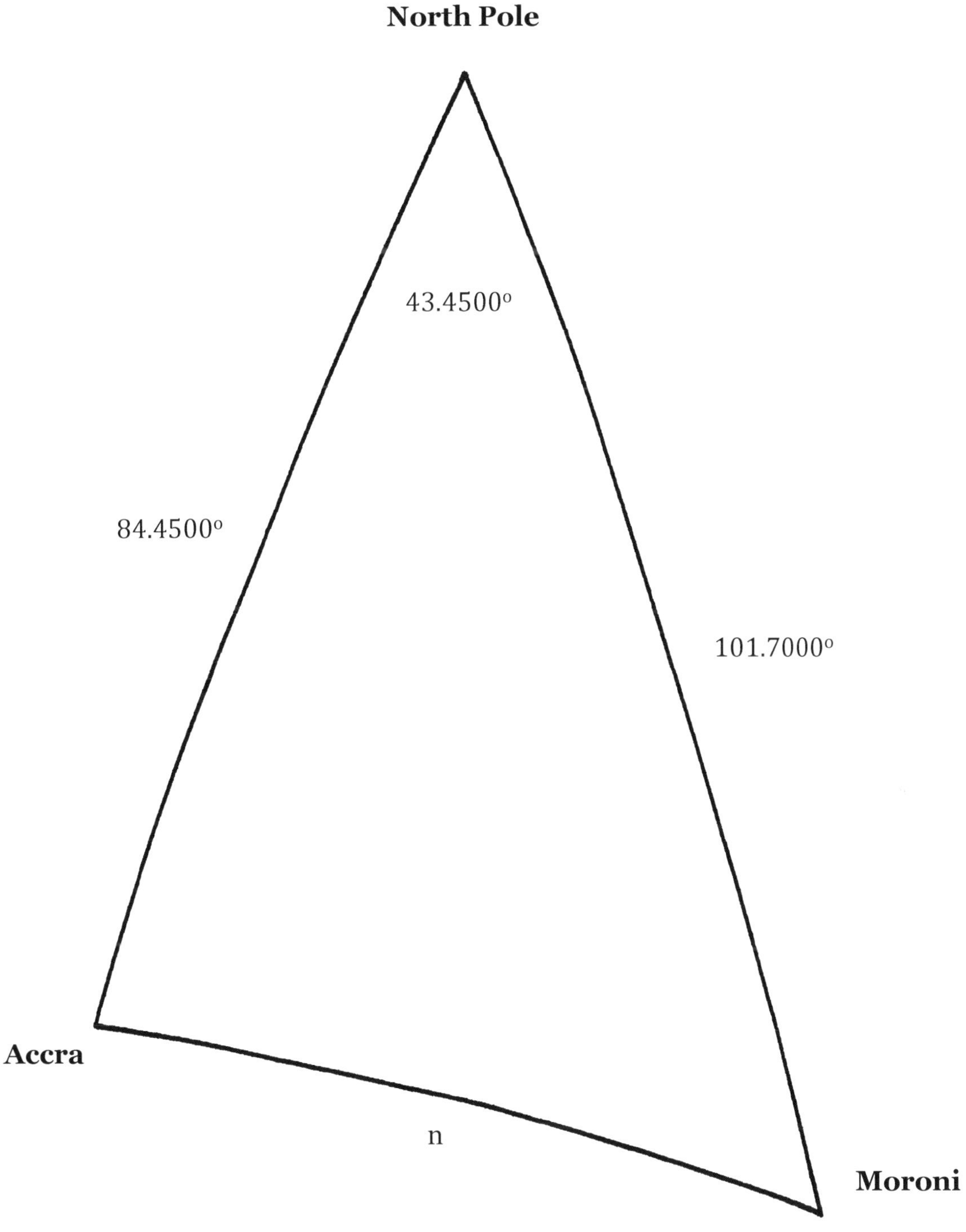
North Pole
43.4500°
84.4500°
101.7000°
Accra
n
Moroni

Chapter 2
Problem 12
Solution

Conakry, Guinea is located at $9^{o}30'\ \mathrm{N} = \left(9 + {}^{30}/_{60}\right)^{o}\ \mathrm{N} = 9.5000^{o}\ \mathrm{N}$,
$13^{o}42'\ \mathrm{W} = \left(13 + {}^{42}/_{60}\right)^{o}\ \mathrm{W} = 13.7000^{o}\ \mathrm{W}$.

Abuja, Nigeria is located at $9^{o}5'\ \mathrm{N} = \left(9 + {}^{5}/_{60}\right)^{o}\ \mathrm{N} \approx 9.0833^{o}\ \mathrm{N}$,
$7^{o}32'\ \mathrm{E} = \left(7 + {}^{32}/_{60}\right)^{o}\ \mathrm{E} \approx 7.5333^{o}\ \mathrm{E}$.

Two sides of the spherical triangle are $90^{o} - 9.5000^{o} = 80.5000^{o}$ and $90^{o} - 9.0833^{o} = 80.9167^{o}$, respectively. The angle at the North Pole is $13.7000^{o} + 7.5333^{o} = 21.2333^{o}$.

See the illustration on the next page.

It follows that

$$\cos n = \cos(80.5000^{o})\cos(80.9167^{o}) + \sin(80.5000^{o})\ \sin(80.9167^{o})\cos(21.2333^{o})$$

$$\cos n \approx 0.933857677061 \Rightarrow n \approx \cos^{-1}(0.933857677061) \Rightarrow n \approx 20.9556284849^{o}.$$

Therefore, the approximate distance between the two capital cities is

$$(20.9556284849^{o})\left({}^{\pi}/_{180^{o}}\right)(3959\ \text{miles})$$

$$\approx (0.365744713886)(3959\ \text{miles})$$

$$\approx 1{,}448\ \text{miles}.$$

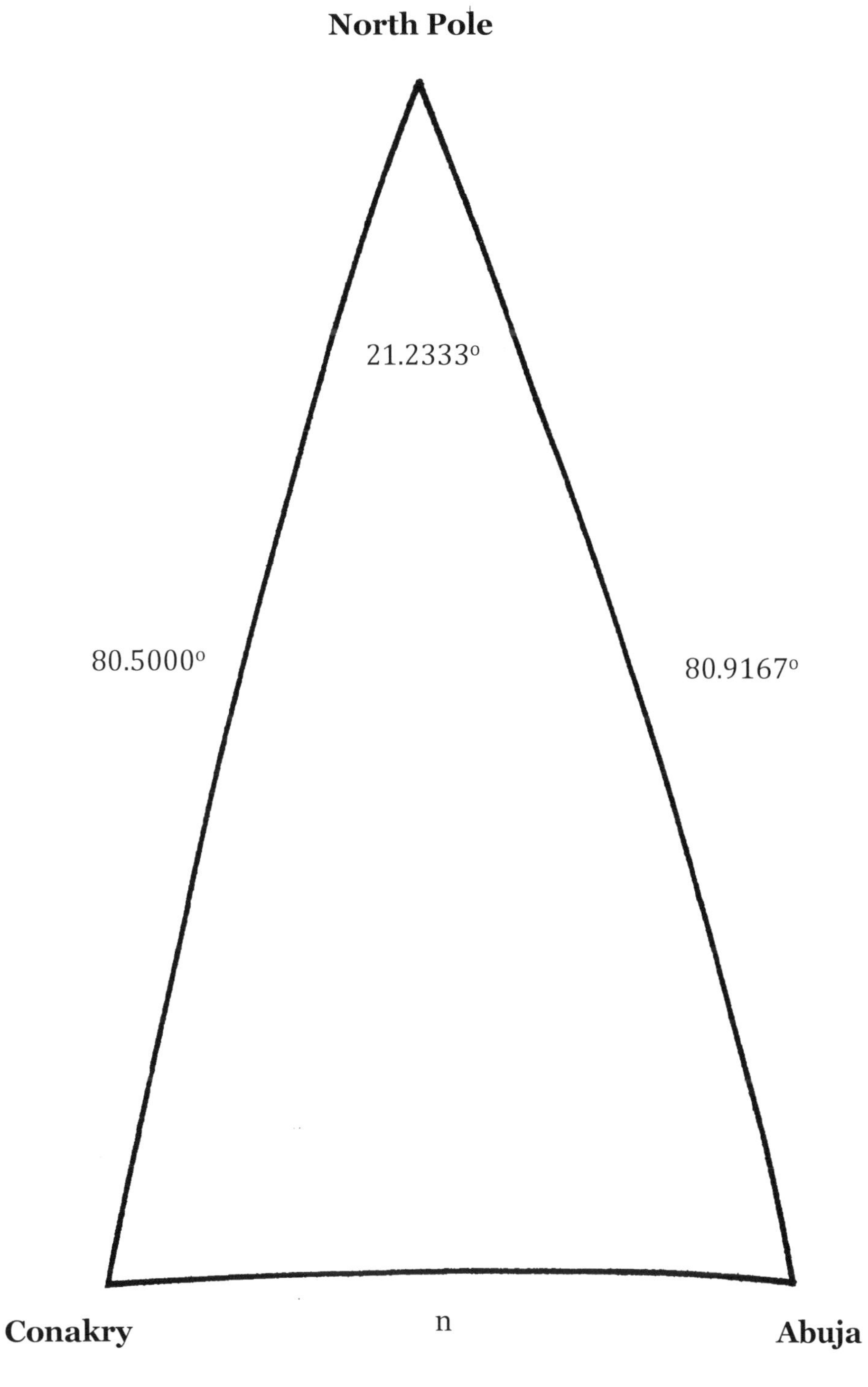
North Pole
21.2333°
80.5000°
80.9167°
Conakry
n
Abuja

Chapter 2
Problem 13
Solution

Maseru, Lesotho is located at $29^{o}19' \text{ S} = \left(29 + {}^{19}/_{60}\right)^{o} \text{ S} \approx 29.3167^{o} \text{ S}$,

$27^{o}29' \text{ E} = \left(27 + {}^{29}/_{60}\right)^{o} \text{ E} \approx 27.4833^{o} \text{ E}$.

Lilongwe, Malawi is located at $13^{o}58' \text{ S} = \left(13 + {}^{58}/_{60}\right)^{o} \text{ S} \approx 13.9667^{o} \text{ S}$,

$33^{o}47' \text{ E} = \left(33 + {}^{47}/_{60}\right)^{o} \text{ E} \approx 33.7833^{o} \text{ E}$.

The sides of the spherical triangle are $90^{o} - 29.3167^{o} = 60.6833^{o}$ and $90^{o} - 13.9667^{o} = 76.0333^{o}$, respectively. The angle at the South Pole is $33.7833^{o} - 27.4833^{o} = 6.3000^{o}$.

See the illustration on the next page.

It follows that

$\cos n = \cos(60.6833^{o})\cos(76.0333^{o}) + \sin(60.6833^{o})\ \sin(76.0333^{o})\cos(6.3000^{o})$

$\cos n \approx 0.959216846415 \ \Rightarrow\ n \approx \cos^{-1}(0.959216846415) \ \Rightarrow\ n \approx 16.4196987834^{o}$.

Therefore, the approximate distance between the two capital cities is

$(16.4196987834^{o})\left({}^{\pi}/_{180^{o}}\right)(3959 \text{ miles})$

$\approx (0.286577805955)(3959 \text{ miles})$

$\approx\ 1{,}135\ \text{ miles}$.

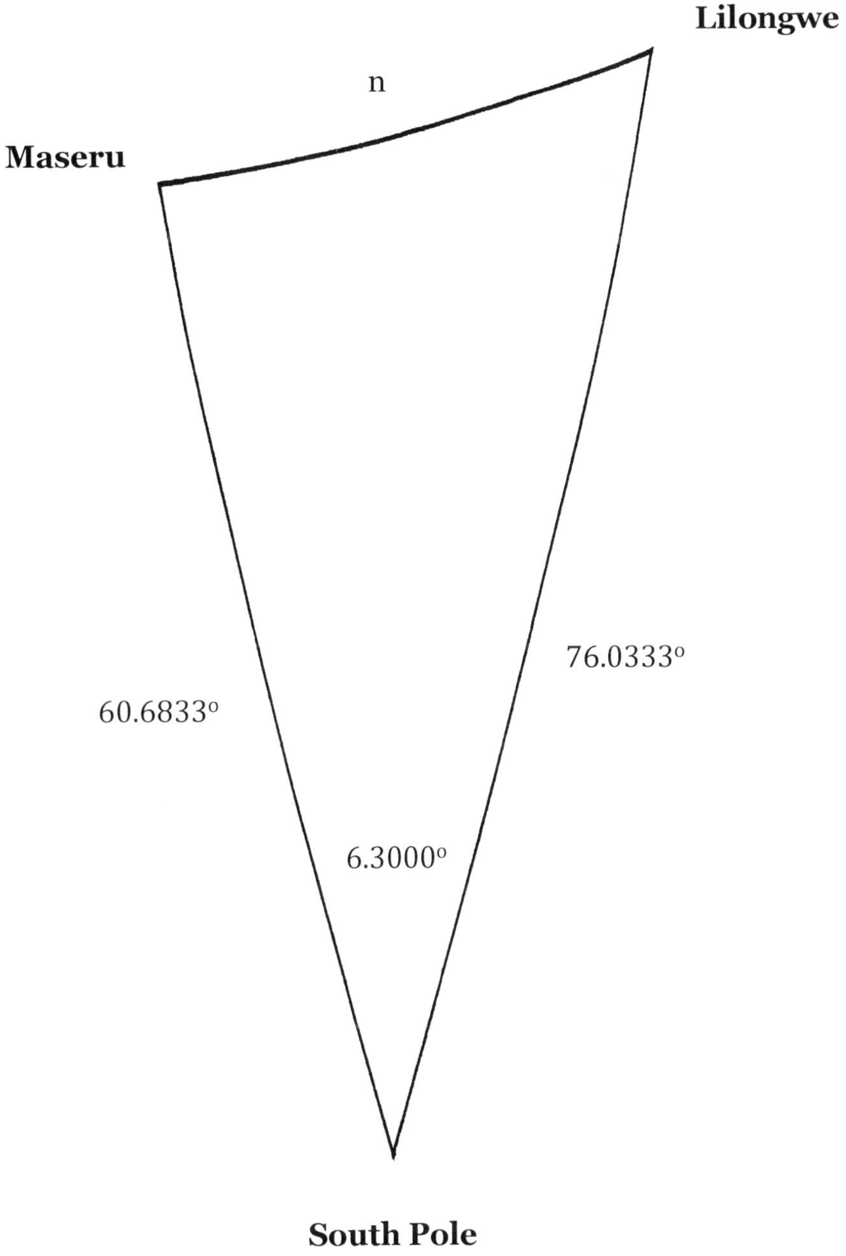
Lilongwe
n
Maseru
76.0333°
60.6833°
6.3000°
South Pole

Chapter 2

Problem 14

Solution

Monrovia, Liberia is located at $6^o18' \text{ N} = \left(6 + {}^{18}/_{60}\right)^o \text{ N} = 6.3000^o \text{ N}$,

$10^o48' \text{ W} = \left(10 + {}^{48}/_{60}\right)^o \text{ W} = 10.8000^o \text{ W}$.

Khartoum, Sudan is located at $15^o36' \text{ N} = \left(15 + {}^{36}/_{60}\right)^o \text{ N} = 15.6000^o \text{ N}$,

$32^o32' \text{ E} = \left(32 + {}^{32}/_{60}\right)^o \text{ E} \approx 32.5333^o \text{ E}$.

Two sides of the spherical triangle are $90^o - 6.3000^o = 83.7000^o$ and $90^o - 15.6000^o = 74.4000^o$, respectively. The angle at the North Pole is $10.8000^o + 32.5333^o = 43.3333^o$.

See the illustration on the next page.

It follows that

$\cos n = \cos(83.7000^o)\cos(74.4000^o) + \sin(83.7000^o)\ \sin(74.4000^o)\cos(43.3333^o)$

$\cos n \approx 0.72585834893 \Rightarrow n \approx \cos^{-1}(0.72585834893) \Rightarrow n \approx 43.4597007479^o$.

Therefore, the approximate distance between the two capital cities is
$(43.4597007479^o)\left({}^{\pi}/_{180^o}\right)(3959 \text{ miles})$

$\approx (0.758514869982)(3959 \text{ miles})$

$\approx 3{,}003$ miles.

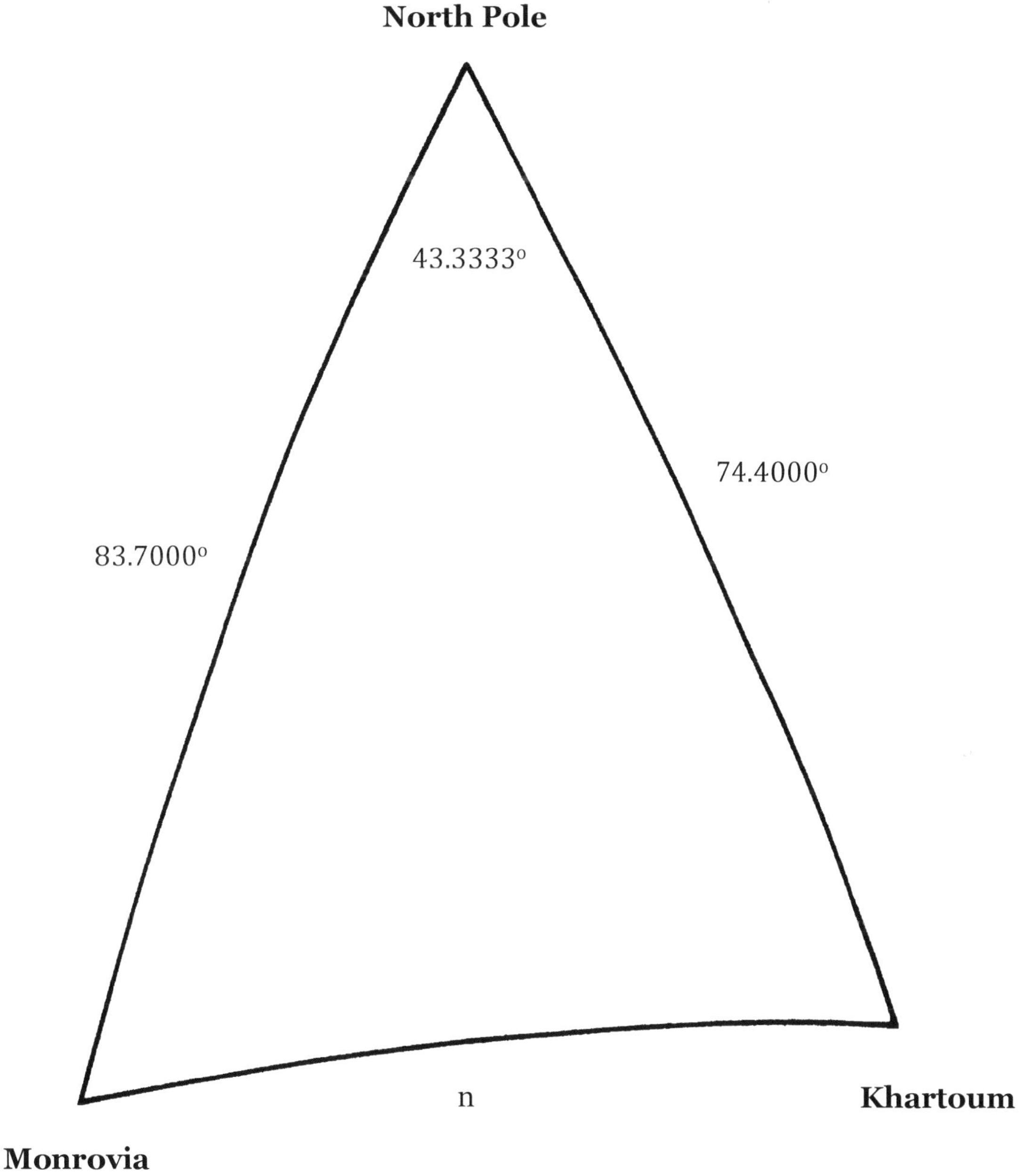
North Pole
43.3333°
74.4000°
83.7000°
n
Khartoum
Monrovia

Chapter 2

Problem 15

Solution

Bamako, Mali is located at $12^o 39' \text{ N} = \left(12 + {}^{39}/_{60}\right)^o \text{ N} = 12.6500^o \text{ N}$,
$8^o 0' \text{ W} = 8^o \text{ W}$.

Bangui, Central African Republic is located at

$4^o 22' \text{ N} = \left(4 + {}^{22}/_{60}\right)^o \text{ N} \approx 4.3667^o \text{ N}$, $18^o 35' \text{ E} = \left(18 + {}^{35}/_{60}\right)^o \text{ E} \approx 18.5833^o \text{ E}$.

Two sides of the spherical triangle are $90^o - 12.6500^o = 77.3500^o$ and $90^o - 4.3667^o = 85.6333^o$, respectively. The angle at the North Pole is $8^o + 18.5833^o = 26.5833^o$.

See the illustration on the next page.

It follows that

$\cos n = \cos(77.3500^o) \cos(85.6333^o) + \sin(77.3500^o) \sin(85.6333^o) \cos(26.5833^o)$

$\cos n \approx 0.886718084381 \Rightarrow n \approx \cos^{-1}(0.886718084381) \Rightarrow n \approx 27.536303455^o$.

Therefore, the approximate distance between the two capital cities is
$(27.536303455^o)\left({}^{\pi}/_{180^o}\right)(3959 \text{ miles})$

$\approx (0.480599159118)(3959 \text{ miles})$

$\approx 1{,}903$ miles.

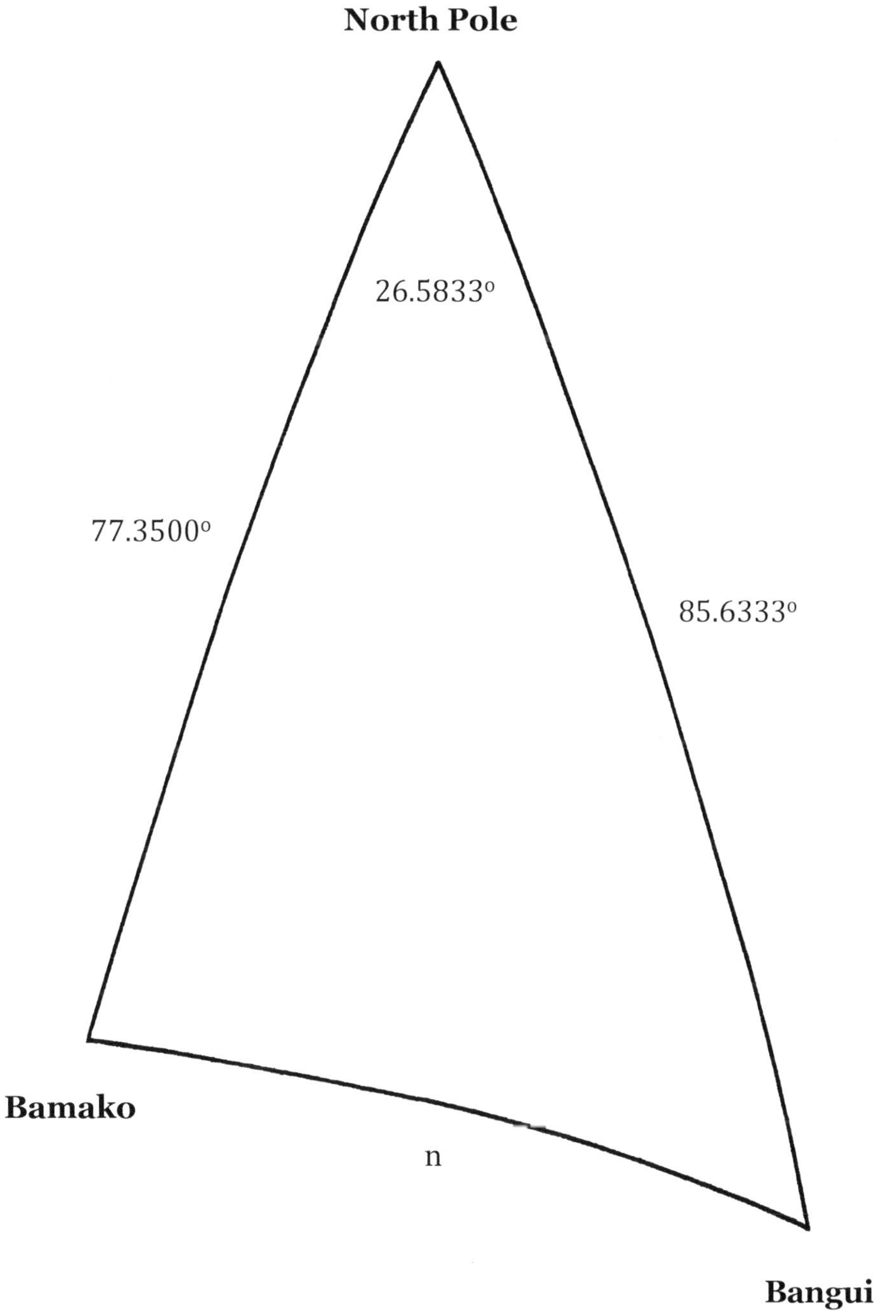
North Pole
26.5833°
77.3500°
85.6333°
Bamako
n
Bangui

Chapter 2

Problem 16

Solution

Nouakchott, Mauritania is located at $18^o4'$ N $= \left(18 + {}^{4}/_{60}\right)^o$ N $\approx 18.0667^o$ N,

$15^o58'$ W $= \left(15 + {}^{58}/_{60}\right)^o$ W $\approx 15.9667^o$ W.

Victoria, Seychelles is located at $4^o37'$ S $= \left(4 + {}^{37}/_{60}\right)^o$ S $\approx 4.6167^o$ S,

$55^o27'$ E $= \left(55 + {}^{27}/_{60}\right)^o$ E $= 55.45^o$ E.

Two sides of the spherical triangle are $90^o - 18.0667^o = 71.9333^o$ and $90^o + 4.6167^o = 94.6167^o$, respectively. The angle at the North Pole is $15.9667^o + 55.45^o = 71.4167^o$.

See the illustration on the next page.

It follows that

$\cos n = \cos(71.9333^o)\cos(94.6167^o) + \sin(71.9333^o)\ \sin(94.6167^o)\cos(71.4167^o)$

$\cos n \approx 0.277026036773 \ \Rightarrow\ n \approx \cos^{-1}(0.277026036773) \ \Rightarrow\ n \approx 73.9172108151^o.$

Therefore, the approximate distance between the two capital cities is

$(73.9172108151^o)\left({}^{\pi}/_{180^o}\right)(3959 \text{ miles})$

$\approx (1.29009870262)(3959 \text{ miles})$

$\approx\ 5{,}108\ \text{ miles}.$

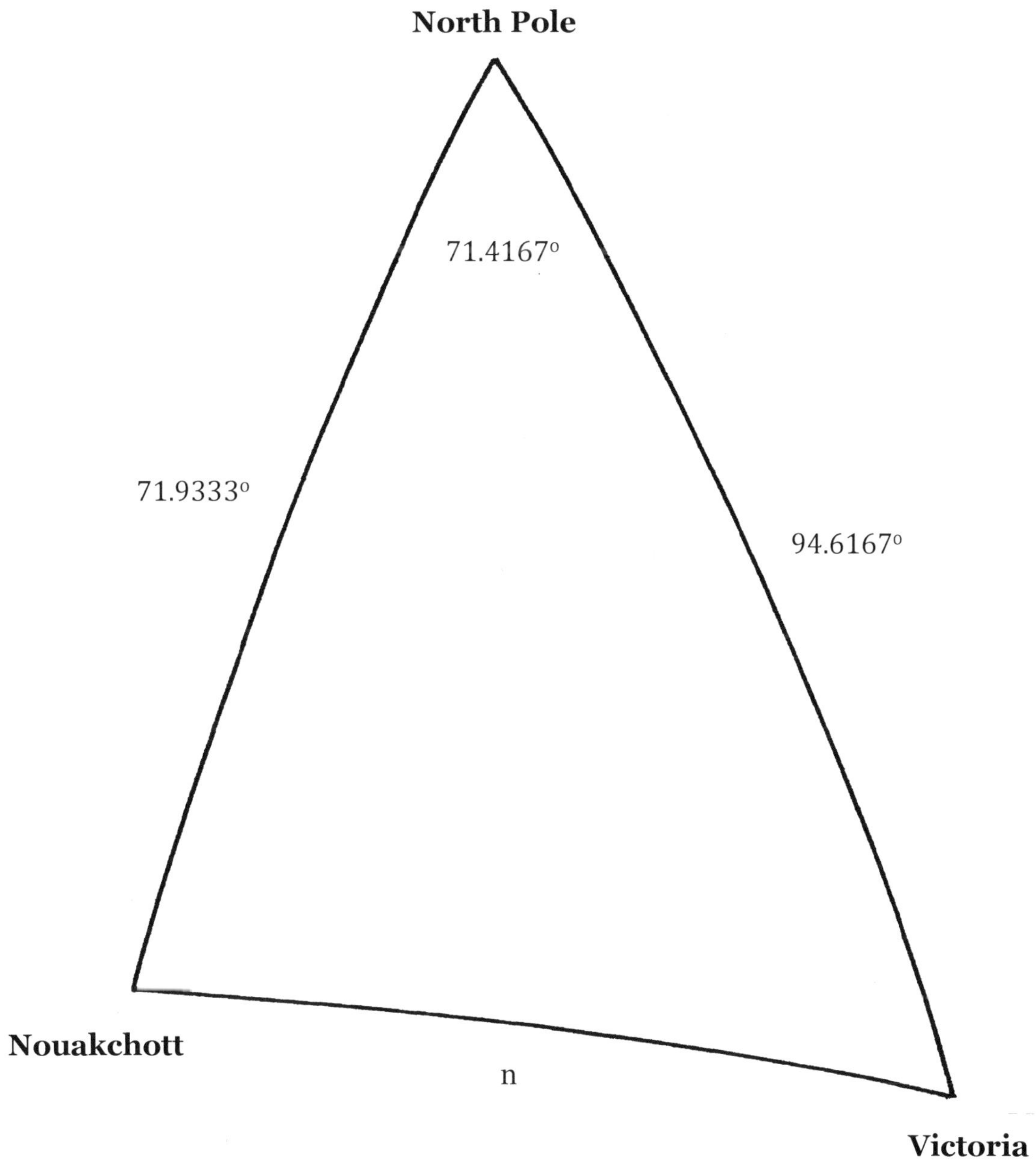
North Pole
71.4167°
71.9333°
94.6167°
Nouakchott
n
Victoria

Chapter 2

Problem 17

Solution

Rabat, Morocco is located at $34^o 1' \text{ N} = \left(34 + {}^{1}/_{60}\right)^o \text{ N} \approx 34.0167^o \text{ N}$,

$6^o 49' \text{ W} = \left(6 + {}^{49}/_{60}\right)^o \text{ W} \approx 6.8167^o \text{ W}$.

Windhoek, Namibia is located at $22^o 34' \text{ S} = \left(22 + {}^{34}/_{60}\right)^o \text{ S} \approx 22.5667^o \text{ S}$,

$17^o 5' \text{ E} = \left(17 + {}^{5}/_{60}\right)^o \text{ E} \approx 17.0833^o \text{ E}$.

Two sides of the spherical triangle are $90^o - 34.0167^o = 55.9833^o$ and $90^o + 22.5667^o = 112.5667^o$, respectively. The angle at the North Pole is $6.8167^o + 17.0833^o = 23.9000^o$.

See the illustration on the next page.

It follows that

$\cos n = \cos(55.9833^o)\cos(112.5667^o) + \sin(55.9833^o)\ \sin(112.5667^o)\cos(23.9000^o)$

$\cos n \approx 0.485091674029 \Rightarrow n \approx \cos^{-1}(0.485091674029) \Rightarrow n \approx 60.9815211844^o$.

Therefore, the approximate distance between the two capital cities is

$(60.9815211844^o)\left({}^{\pi}/_{180^o}\right)(3959 \text{ miles})$

$\approx (1.06432832754)(3959 \text{ miles})$

$\approx 4{,}214$ miles.

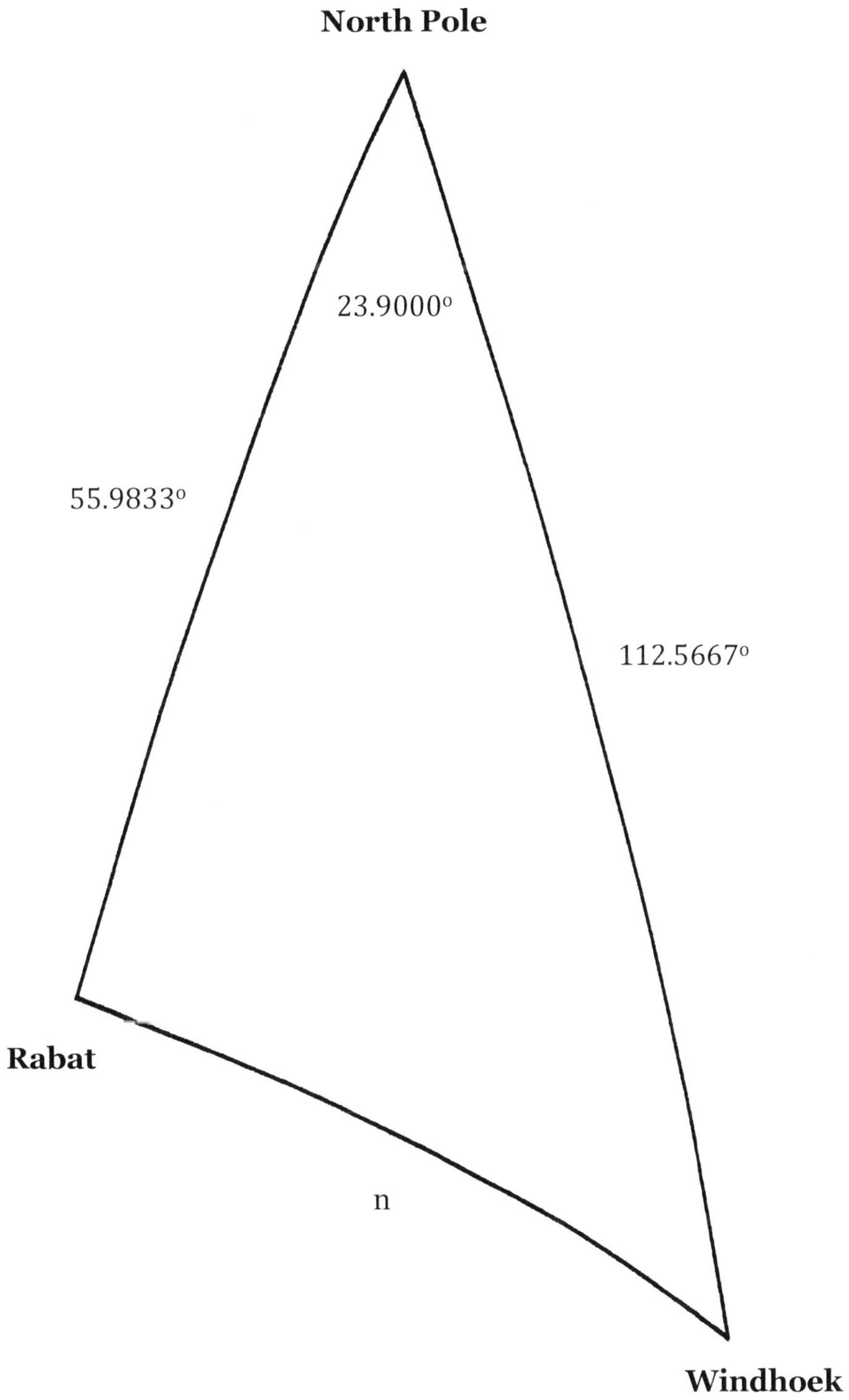
North Pole
23.9000°
55.9833°
112.5667°
Rabat
n
Windhoek

Chapter 2

Problem 18

Solution

Niamey, Niger is located at $13^o31'$ N $= \left(13 + {}^{31}/_{60}\right)^o$ N $\approx 13.5167^o$ N ,

$2^o7'$ E $= \left(2 + {}^{7}/_{60}\right)^o$ E $\approx 2.1167^o$ E .

Kigali, Rwanda is located at $1^o57'$ S $= \left(1 + {}^{57}/_{60}\right)^o$ S $= 1.9500^o$ S ,

$30^o3'$ E $= \left(30 + {}^{3}/_{60}\right)^o$ E $= 30.0500^o$ E .

Two sides of the spherical triangle are $90^o - 13.5167^o = 76.4833^o$ and $90^o + 1.9500^o = 91.9500^o$, respectively. The angle at the North Pole is $30.0500^o - 2.1167^o = 27.9333^o$.

See the illustration on the next page.

It follows that

$\cos n = \cos(76.4833^o)\cos(91.9500^o) + \sin(76.4833^o)\ \sin(91.9500^o)\cos(27.9333^o)$

$\cos n \approx 0.850571745838 \Rightarrow n \approx \cos^{-1}(0.850571745838) \Rightarrow n \approx 31.7260898373^o.$

Therefore, the approximate distance between the two capital cities is

$(31.7260898373^o)\left({}^{\pi}/_{180^o}\right)(3959 \text{ miles})$

$\approx (0.553724726445)(3959 \text{ miles})$

$\approx$ 2,192 miles.

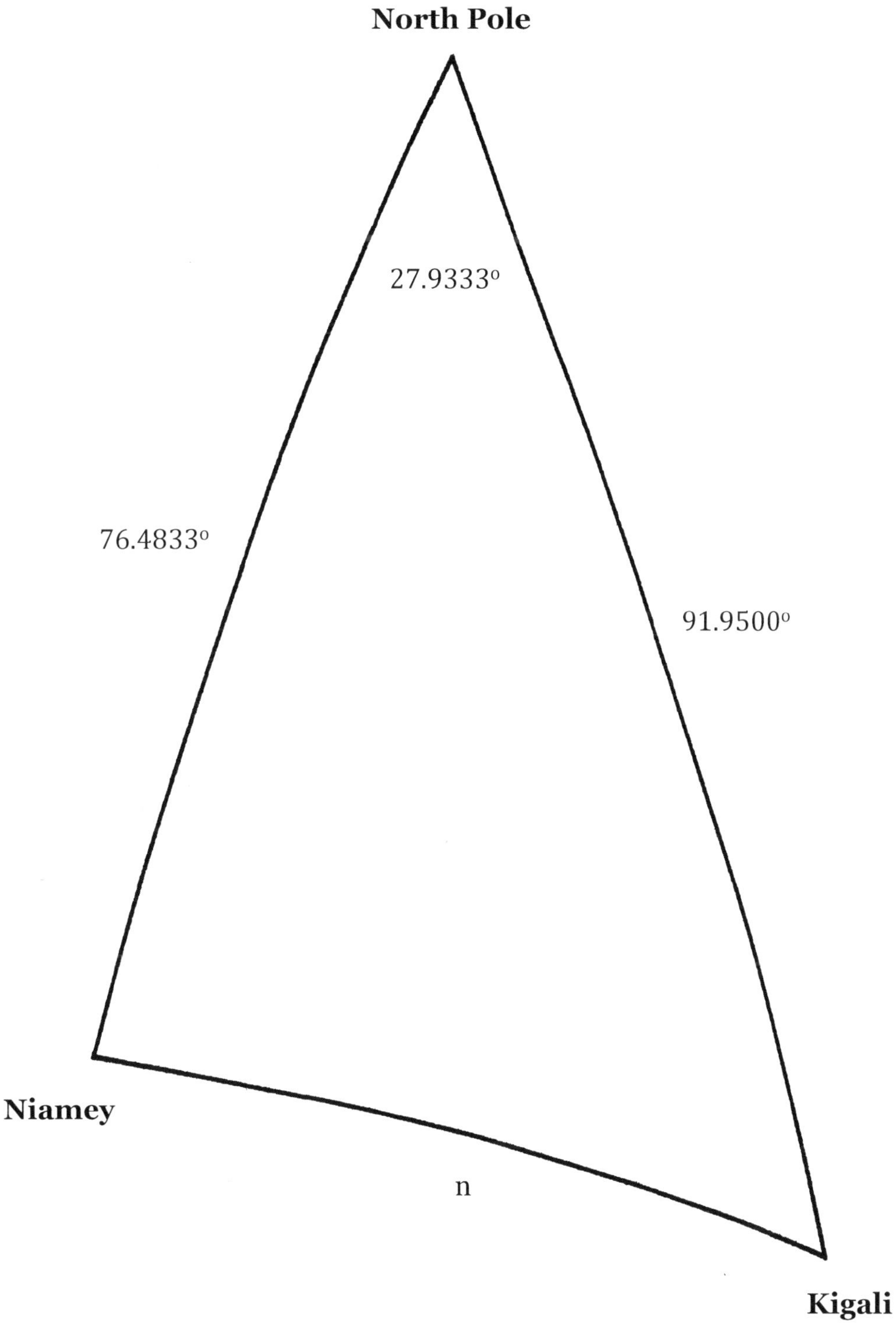
North Pole
27.9333°
76.4833°
91.9500°
Niamey
n
Kigali

Chapter 2

Problem 19

Solution

Brazzaville, Republic of the Congo is located at

$4^o15'\ S = \left(4 + {}^{15}/_{60}\right)^o S = 4.2500^o\ S\ , 15^o17'\ E = \left(15 + {}^{17}/_{60}\right)^o E \approx 15.2833^o\ E\ .$

Djibouti, Djibouti is located at $11^o35'\ N = \left(11 + {}^{35}/_{60}\right)^o N \approx 11.5833^o\ N\ ,$

$43^o9'\ E = \left(43 + {}^{9}/_{60}\right)^o E = 43.1500^o\ E\ .$

Two sides of the spherical triangle are $90^o + 4.2500^o = 94.2500^o$ and $90^o - 11.5833^o = 78.4167^o$, respectively. The angle at the North Pole is $43.1500^o - 15.2833^o = 27.8667^o$.

See the illustration on the next page.

It follows that

$\cos n = \cos(94.2500^o)\cos(78.4167^o) + \sin(94.2500^o)\ \sin(78.4167^o)\cos(27.8667^o)$

$\cos n \approx 0.848771118449 \quad \Rightarrow\ n \approx \cos^{-1}(0.848771118449\ ) \Rightarrow\ n \approx 31.9217400319^o.$

Therefore, the approximate distance between the two capital cities is

$(31.9217400319^o)\left({}^{\pi}/_{180^o}\right)(3959 \text{ miles})$

$\approx (0.557139466523)(3959 \text{ miles})$

$\approx\ 2{,}206 \text{ miles.}$

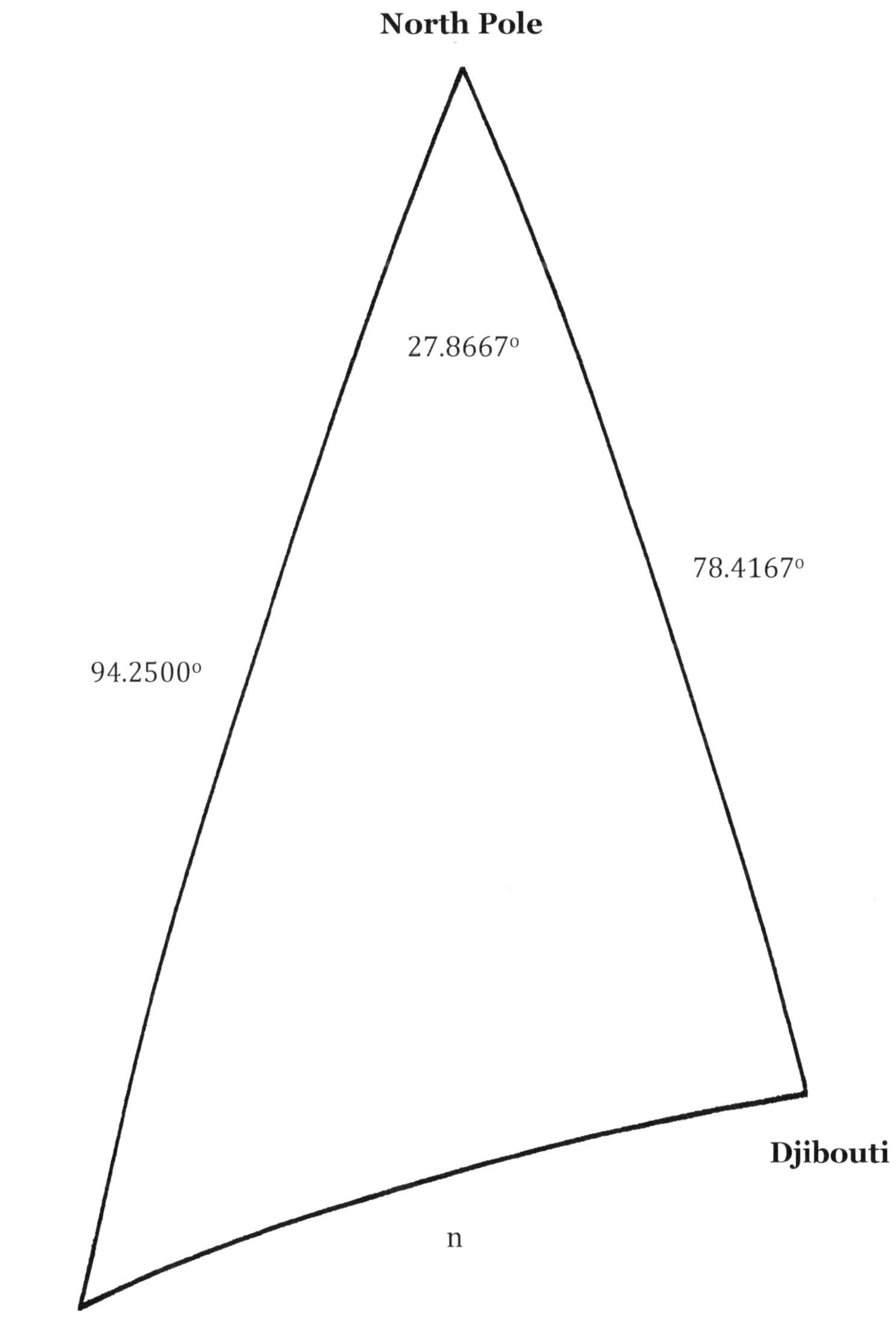
North Pole
27.8667°
78.4167°
94.2500°
Djibouti
n
Brazzaville

Chapter 2

Problem 20

Solution

Dakar, Senegal is located at $14^\circ 44'\ \mathrm{N} = \left(14 + {}^{44}/_{60}\right)^\circ \mathrm{N} \approx 14.7333^\circ\ \mathrm{N}$,

$17^\circ 38'\ \mathrm{W} = \left(17 + {}^{38}/_{60}\right)^\circ \mathrm{W} \approx 17.6333^\circ\ \mathrm{W}$.

Cairo, Egypt is located at $30^\circ 3'\ \mathrm{N} = \left(30 + {}^{3}/_{60}\right)^\circ \mathrm{N} = 30.0500^\circ\ \mathrm{N}$,

$31^\circ 15'\ \mathrm{E} = \left(31 + {}^{15}/_{60}\right)^\circ \mathrm{E} = 31.2500^\circ\ \mathrm{E}$.

Two sides of the spherical triangle are $90^\circ - 14.7333^\circ = 75.2667^\circ$ and $90^\circ - 30.0500^\circ = 59.9500^\circ$, respectively. The angle at the North Pole is $17.6333^\circ + 31.2500^\circ = 48.8833^\circ$.

See the illustration on the next page.

It follows that

$\cos n = \cos(75.2667^\circ)\cos(59.9500^\circ) + \sin(75.2667^\circ)\ \sin(59.9500^\circ)\cos(48.8833^\circ)$

$\cos n \approx 0.677843440925 \ \Rightarrow\ n \approx \cos^{-1}(0.677843440925) \ \Rightarrow\ n \approx 47.3246491072^\circ$.

Therefore, the approximate distance between the two capital cities is

$(47.3246491072^\circ)\left({}^{\pi}/_{180^\circ}\right)(3959 \text{ miles})$

$\approx (0.825970944272)(3959 \text{ miles})$

$\approx\ 3{,}270\ \text{miles}$.

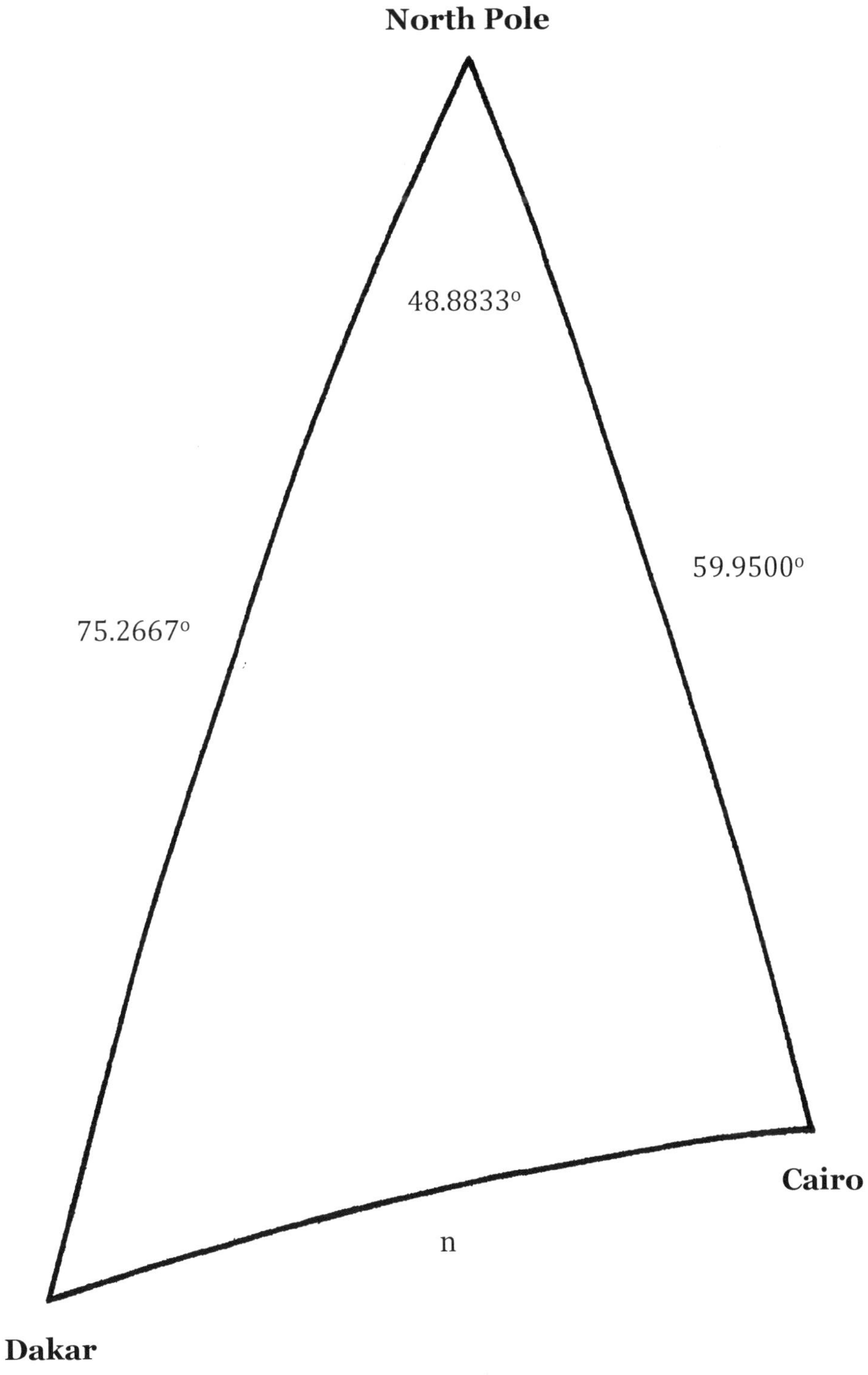
North Pole
48.8833°
59.9500°
75.2667°
Cairo
n
Dakar

Chapter 2
Problem 21
Solution

Freetown, Sierra Leone is located at $8^o29' \text{ N} = \left(8 + {}^{29}/_{60}\right)^o \text{ N} \approx 8.4833^o \text{ N}$,

$13^o14' \text{ W} = \left(13 + {}^{14}/_{60}\right)^o \text{ W} \approx 13.2333^o \text{ W}$.

Bujumbura, Burundi is located at $3^o22' \text{ S} = \left(3 + {}^{22}/_{60}\right)^o \text{ S} \approx 3.3667^o \text{ S}$,

$29^o21' \text{ E} = \left(29 + {}^{21}/_{60}\right)^o \text{ E} = 29.3500^o \text{ E}$.

Two sides of the spherical triangle are $90^o - 8.4833^o = 81.5167^o$ and $90^o + 3.3667^o = 93.3667^o$, respectively. The angle at the North Pole is $13.2333^o + 29.3500^o = 42.5833^o$.

See the illustration on the next page.

It follows that

$\cos n = \cos(81.5167^o)\cos(93.3667^o) + \sin(81.5167^o)\ \sin(93.3667^o)\cos(42.5833^o)$

$\cos n \approx 0.718318276229 \Rightarrow n \approx \cos^{-1}(0.718318276229\ ) \Rightarrow n \approx 44.0841918095^o$.

Therefore, the approximate distance between the two capital cities is

$(44.0841918095^o)\left({}^{\pi}/_{180^o}\right)(3959 \text{ miles})$

$\approx (0.769414295157)(3959 \text{ miles})$

$\approx 3{,}046$ miles.

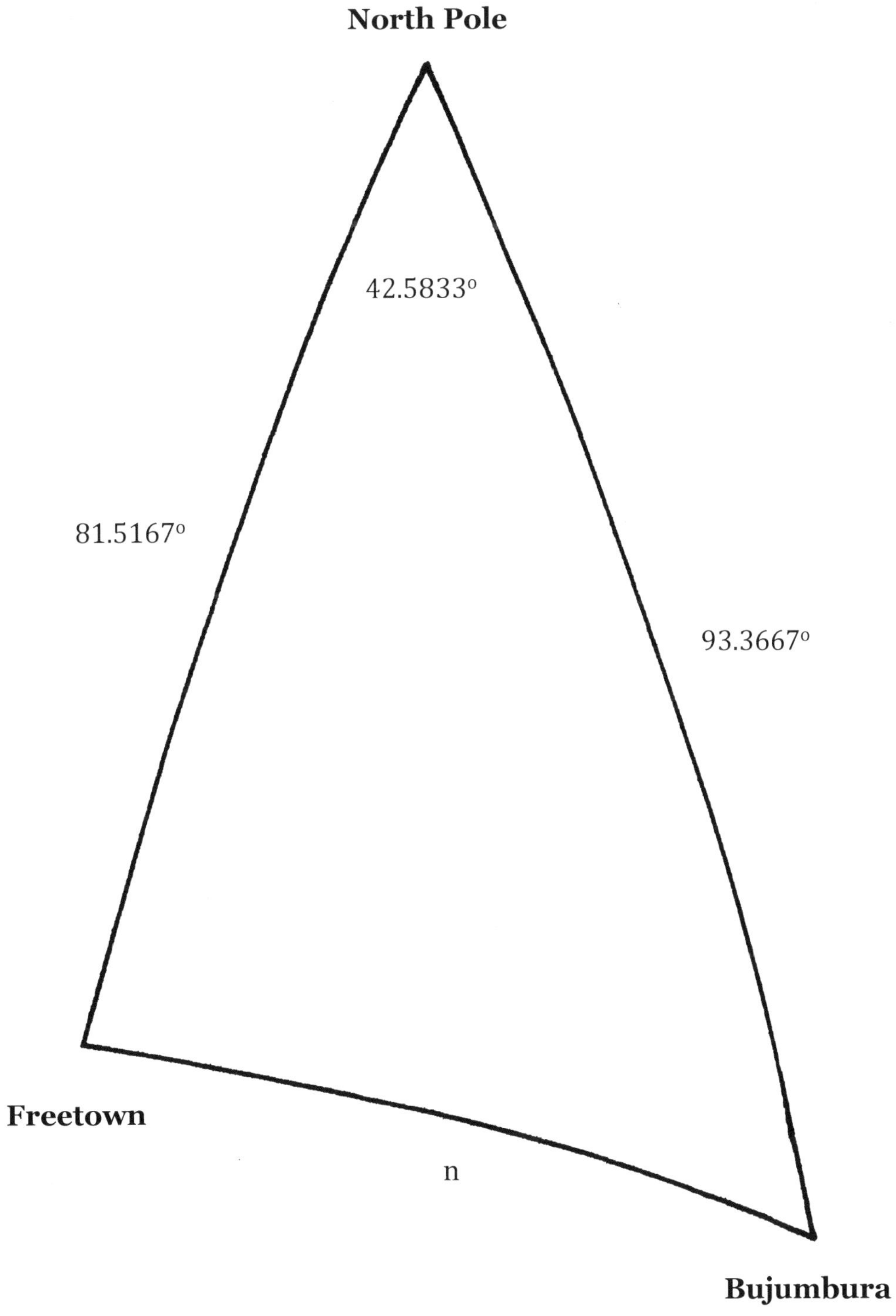
North Pole
42.5833°
81.5167°
93.3667°
Freetown
n
Bujumbura

Chapter 2
Problem 22
Solution

Juba, South Sudan is located at $4^o51' \text{ N} = \left(4 + {}^{51}/_{60}\right)^o \text{ N} = 4.8500^o \text{ N}$,

$31^o37' \text{ E} = \left(31 + {}^{37}/_{60}\right)^o \text{ E} \approx 31.6167^o \text{ E}$.

Mogadishu, Somalia is located at $2^o4' \text{ N} = \left(2 + {}^{4}/_{60}\right)^o \text{ N} \approx 2.0667^o \text{ N}$,

$45^o20' \text{ E} = \left(45 + {}^{20}/_{60}\right)^o \text{ E} \approx 45.3333^o \text{ E}$.

Two sides of the spherical triangle are $90^o - 4.8500^o = 85.1500^o$ and $90^o - 2.0667^o = 87.9333^o$, respectively. The angle at the North Pole is $45.3333^o - 31.6167^o = 13.7166^o$.

See the illustration on the next page.

It follows that

$\cos \text{n} = \cos(85.1500^o)\cos(87.9333^o) + \sin(85.1500^o)\ \sin(87.9333^o)\cos(13.7166^o)$

$\cos \text{n} \approx 0.970421394864 \ \Rightarrow\ \text{n} \approx \cos^{-1}(0.970421394864) \ \Rightarrow\ \text{n} \approx 13.9702061016^o$.

Therefore, the approximate distance between the two capital cities is

$(13.9702061016^o)\left({}^{\pi}/_{180^o}\right)(3959 \text{ miles})$

$\approx (0.243826093655)(3959 \text{ miles})$

$\approx\ 965\ \text{ miles}$.

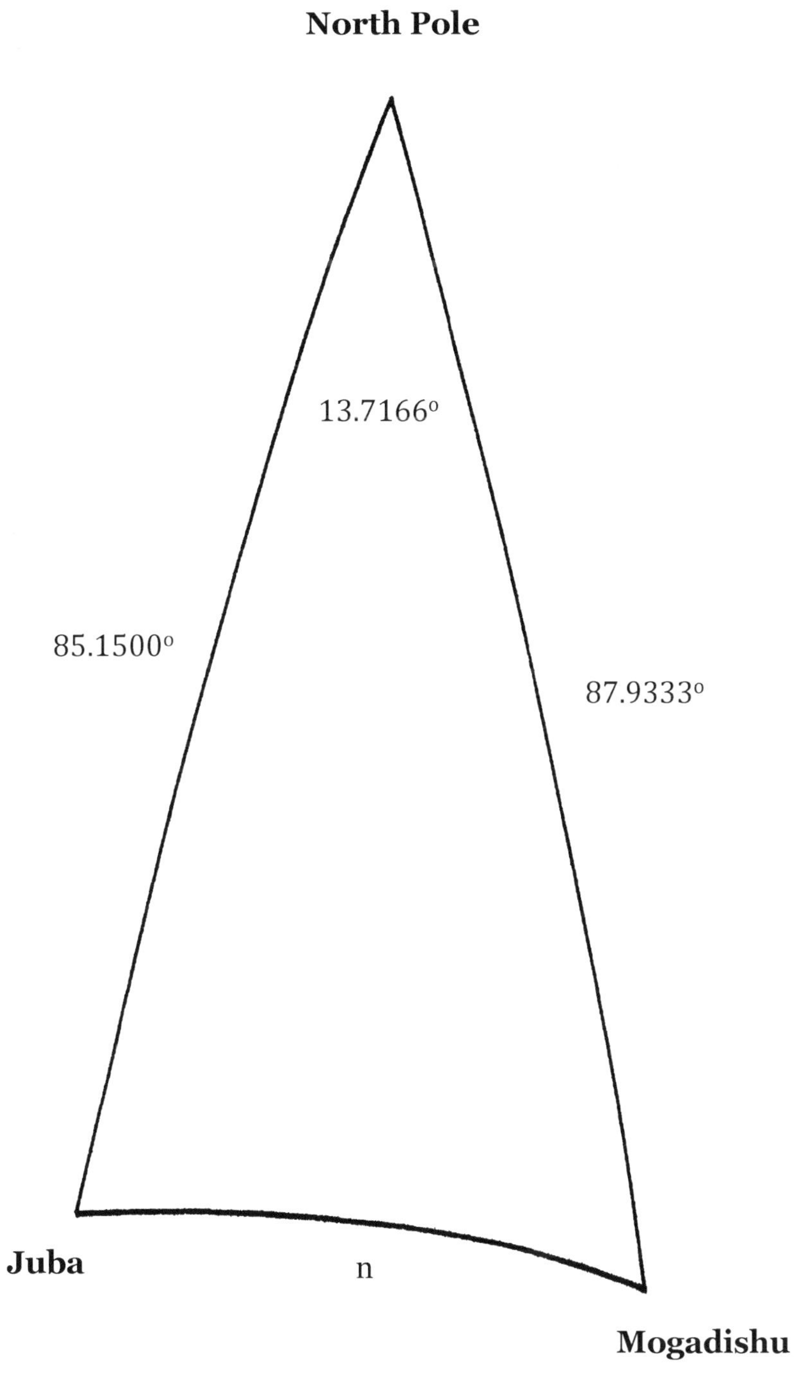
North Pole
13.7166°
85.1500°
87.9333°
Juba
n
Mogadishu

Chapter 2

Problem 23

Solution

Lomé, Togo is located at $6^o7'$ N $= (6 + {}^{7}/_{60})^o$ N $\approx 6.1167^o$ N ,

$1^o13'$ E $= (1 + {}^{13}/_{60})^o$ E $\approx 1.2167^o$ E .

Maputo, Mozambique is located at $25^o57'$ S $= \left(25 + {}^{57}/_{60}\right)^o$ S $= 25.9500^o$ S ,

$32^o35'$E $= \left(32 + {}^{35}/_{60}\right)^o$ E $\approx 32.5833^o$ E .

Two sides of the spherical triangle are $90^o - 6.1167^o = 83.8833^o$ and $90^o + 25.9500^o = 115.9500^o$, respectively. The angle at the North Pole is $32.5833^o - 1.2167^o = 31.3666^o$.

See the illustration on the next page.

It follows that

$\cos n = \cos(83.8833^o)\cos(115.9500^o) + \sin(83.8833^o)\ \sin(115.9500^o)\cos(31.3666^o)$

$\cos n \approx 0.716768073964 \Rightarrow n \approx \cos^{-1}(0.716768073964) \Rightarrow n \approx 44.2117127499^o$.

Therefore, the approximate distance between the two capital cities is

$(44.2117127499^o)({}^{\pi}/_{180^o})(3959 \text{ miles})$

$\approx (0.771639955431)(3959 \text{ miles})$

$\approx$ 3,055 miles.

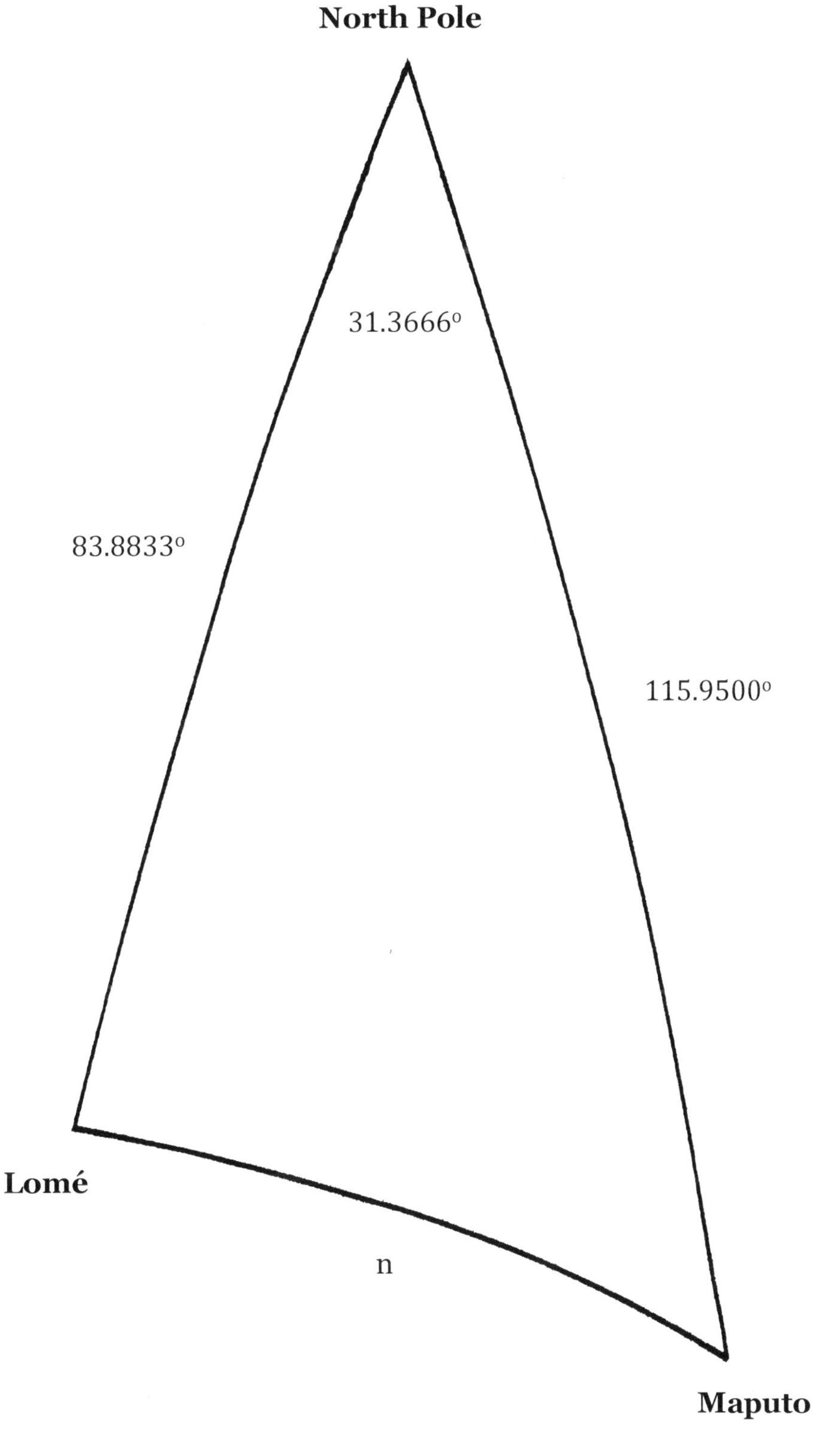
North Pole
31.3666°
83.8833°
115.9500°
Lomé
n
Maputo

Chapter 2
Problem 24
Solution

Tunis, Tunisia is located at $36^o48' \text{ N} = \left(36 + {}^{48}/_{60}\right)^o \text{ N} = 36.8000^o \text{ N}$,
$10^o11' \text{ E} = \left(10 + {}^{11}/_{60}\right)^o \text{ E} \approx 10.1833^o \text{ E}$.

Port Louis, Mauritius is located at $20^o9' \text{ S} = \left(20 + {}^{9}/_{60}\right)^o \text{ S} = 20.1500^o \text{ S}$,
$57^o29' \text{ E} = \left(57 + {}^{29}/_{60}\right)^o \text{ E} \approx 57.4833^o \text{ E}$.

Two sides of the spherical triangle are $90^o - 36.8000^o = 53.2000^o$ and
$90^o + 20.1500^o = 110.1500^o$, respectively. The angle at the North Pole is
$57.4833^o - 10.1833^o = 47.3000^o$.

See the illustration on the next page.

It follows that
$\cos n = \cos(53.2000^o)\cos(110.1500^o) + \sin(53.2000^o)\sin(110.1500^o)\cos(47.3000^o)$
$\cos n \approx 0.3034363126 \Rightarrow n \approx \cos^{-1}(0.3034363126) \Rightarrow n \approx 72.33588656^o$.

Therefore, the approximate distance between the two capital cities is
$(72.33588656^o)\left({}^{\pi}/_{180^o}\right)(3959 \text{ miles})$
$\approx (1.262499388)(3959 \text{ miles})$
≈ 4998 miles.

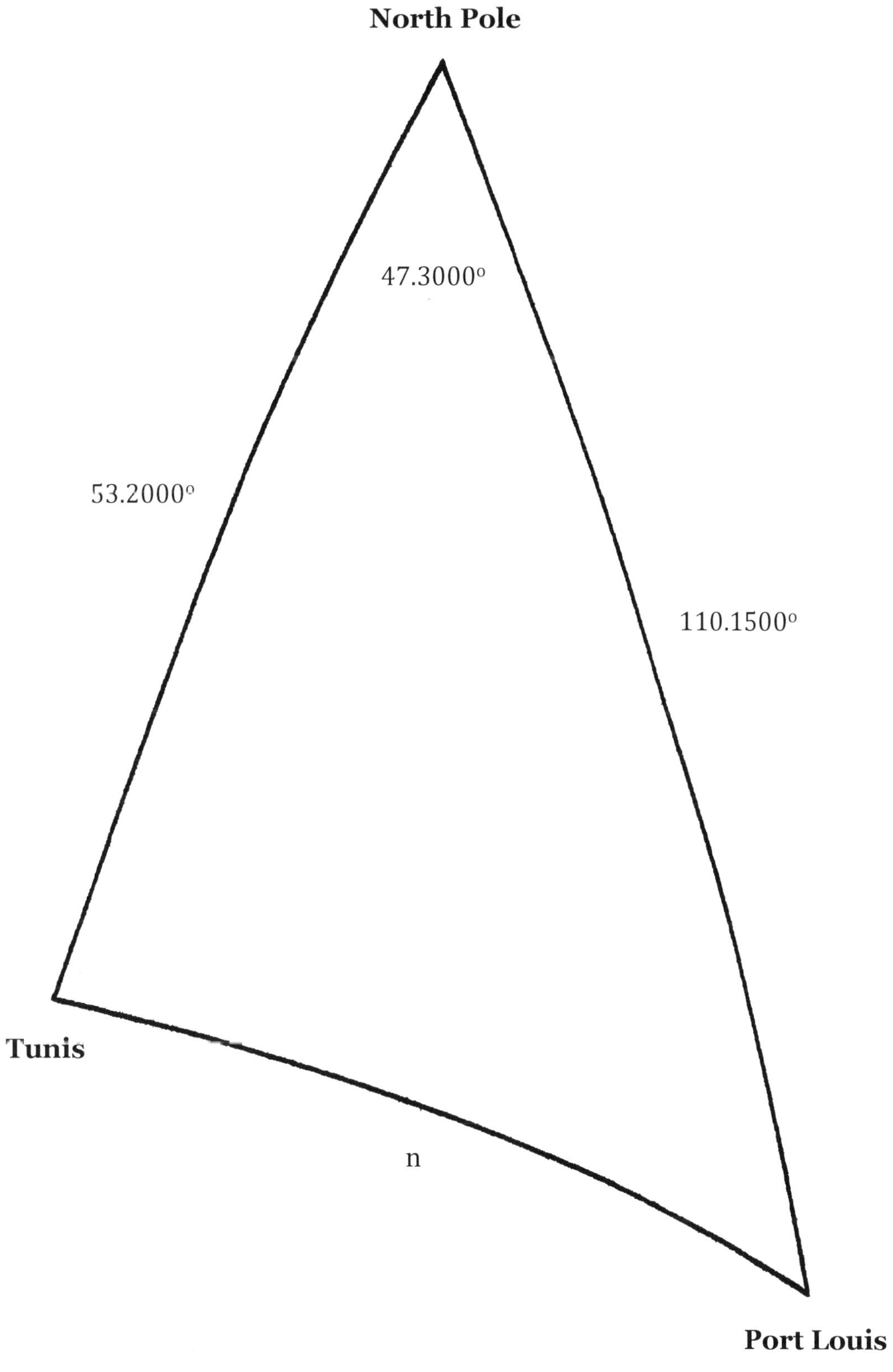
North Pole
47.3000°
53.2000°
110.1500°
Tunis
n
Port Louis

Chapter 3

Australia & Oceania

Example 1

The Settlement, Christmas Island is located at

$10^o25' \text{ S} = \left(10 + {}^{25}/_{60}\right)^o \text{ S} \approx 10.4167^o \text{ S}$,

$105^o43' \text{ E} = \left(105 + {}^{43}/_{60}\right)^o \text{ E} \approx 105.7167^o \text{ E}$.

Nuku'alofa, Tonga is located at $21^o8' \text{ S} = \left(21 + {}^{8}/_{60}\right)^o \text{ S} \approx 21.1333^o \text{ S}$,

$175^o12' \text{ W} = \left(175 + {}^{12}/_{60}\right)^o \text{ W} = 175.2000^o \text{ W}$.

Two sides of the spherical triangle are $90^o - 10.4167^o = 79.5833^o$ and $90^o - 21.1333^o = 68.8667^o$, respectively. The angle at the South Pole is $(180^o - 105.7167^o) + (180^o - 175.2000^o) = 74.2833^o + 4.8000^o = 79.0833^o$.

See the illustration on the next page.

It follows that

$\cos n = \cos(79.5833^o)\cos(68.8667^o) + \sin(79.5833^o)\ \sin(68.8667^o)\cos(79.0833^o)$

$\cos n \approx 0.238920854308 \Rightarrow n \approx \cos^{-1}(0.238920854308) \Rightarrow n \approx 76.1771429285^o$.

Therefore, the approximate distance between the two capital cities is
$(76.1771429285^o)\left({}^{\pi}/_{180^o}\right)(3959 \text{ miles})$

$\approx (1.32954195886)(3959 \text{ miles})$

$\approx$ 5,264 miles.

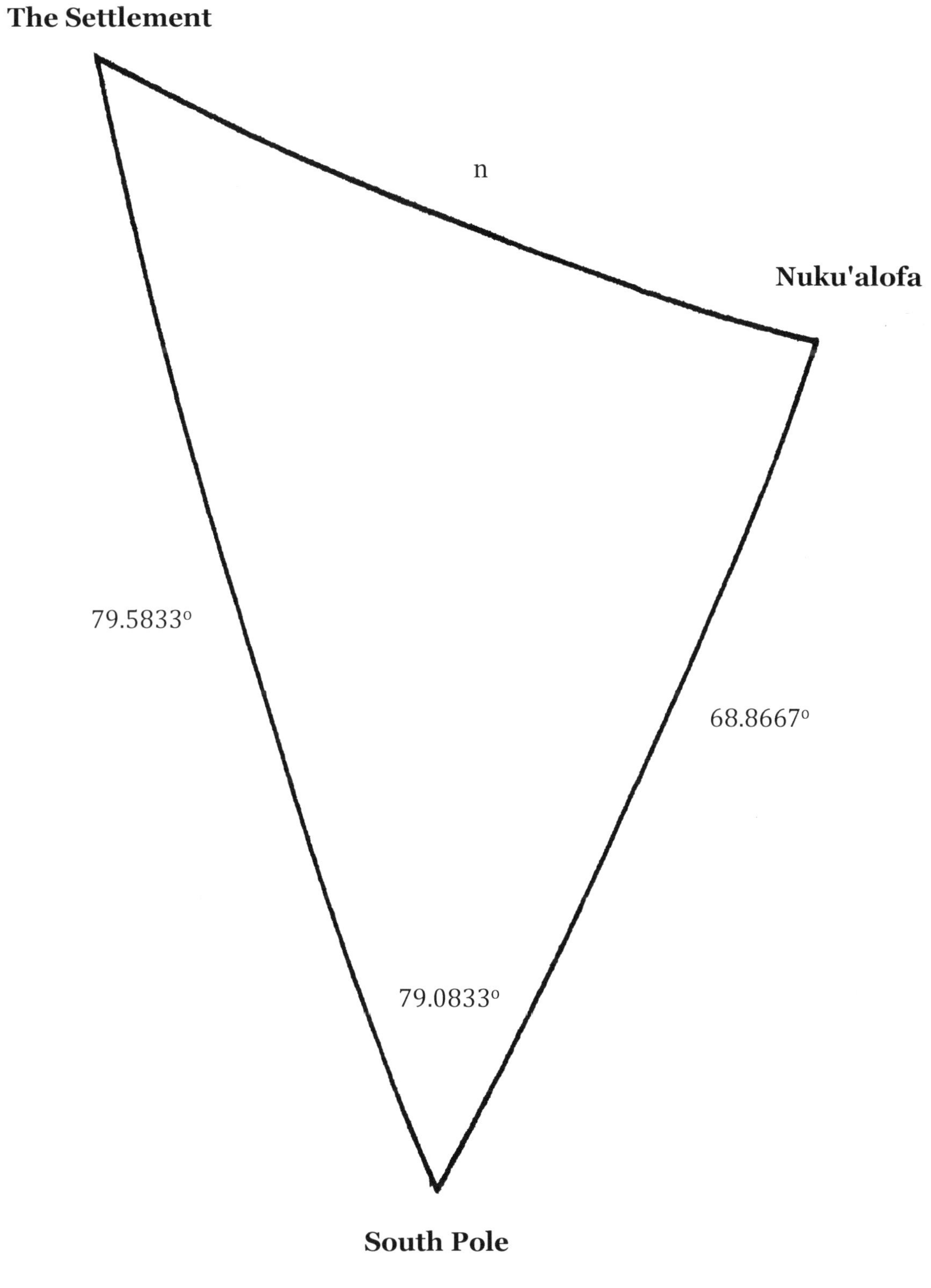

The Settlement
n
Nuku'alofa
79.5833°
68.8667°
79.0833°
South Pole

Example 2

Majuro, Marshall Islands is located at $7^{o}6'$ N $= \left(7 + {}^{6}/_{60}\right)^{o}$ N $= 7.1000^{o}$ N ,

$171^{o}23'$ E $= \left(171 + {}^{23}/_{60}\right)^{o}$ E $\approx 171.3833^{o}$ E .

Adamstown, Pitcairn Islands is located at $25^{o}4'$ S $= \left(25 + {}^{4}/_{60}\right)^{o}$ S $\approx 25.0667^{o}$ S ,

$130^{o}5'$ W $= \left(130 + {}^{5}/_{60}\right)^{o}$ W $\approx 130.0833^{o}$ W .

Two sides of the spherical triangle are $90^{o} + 7.1000^{o} = 97.1000^{o}$ and $90^{o} - 25.0667^{o} = 64.9333^{o}$, respectively. The angle at the South Pole is $(180^{o} - 171.3833^{o}) + (180^{o} - 130.0833^{o}) = 8.6167^{o} + 49.9167^{o} = 58.5334^{o}$.

See the illustration on the next page.

It follows that

$\cos n = \cos(97.1000^{o})\cos(64.9333^{o}) + \sin(97.1000^{o})\ \sin(64.9333^{o})\cos(58.5334^{o})$

$\cos n \approx 0.416844475378 \ \Rightarrow\ n \approx \cos^{-1}(0.416844475378)\ \Rightarrow\ n \approx 65.3644742987^{o}$.

Therefore, the approximate distance between the two capital cities is $(65.3644742987^{o})\left({}^{\pi}/_{180^{o}}\right)(3959 \text{ miles})$

$\approx (1.14082529035)(3959 \text{ miles})$

$\approx$ 4,517 miles.

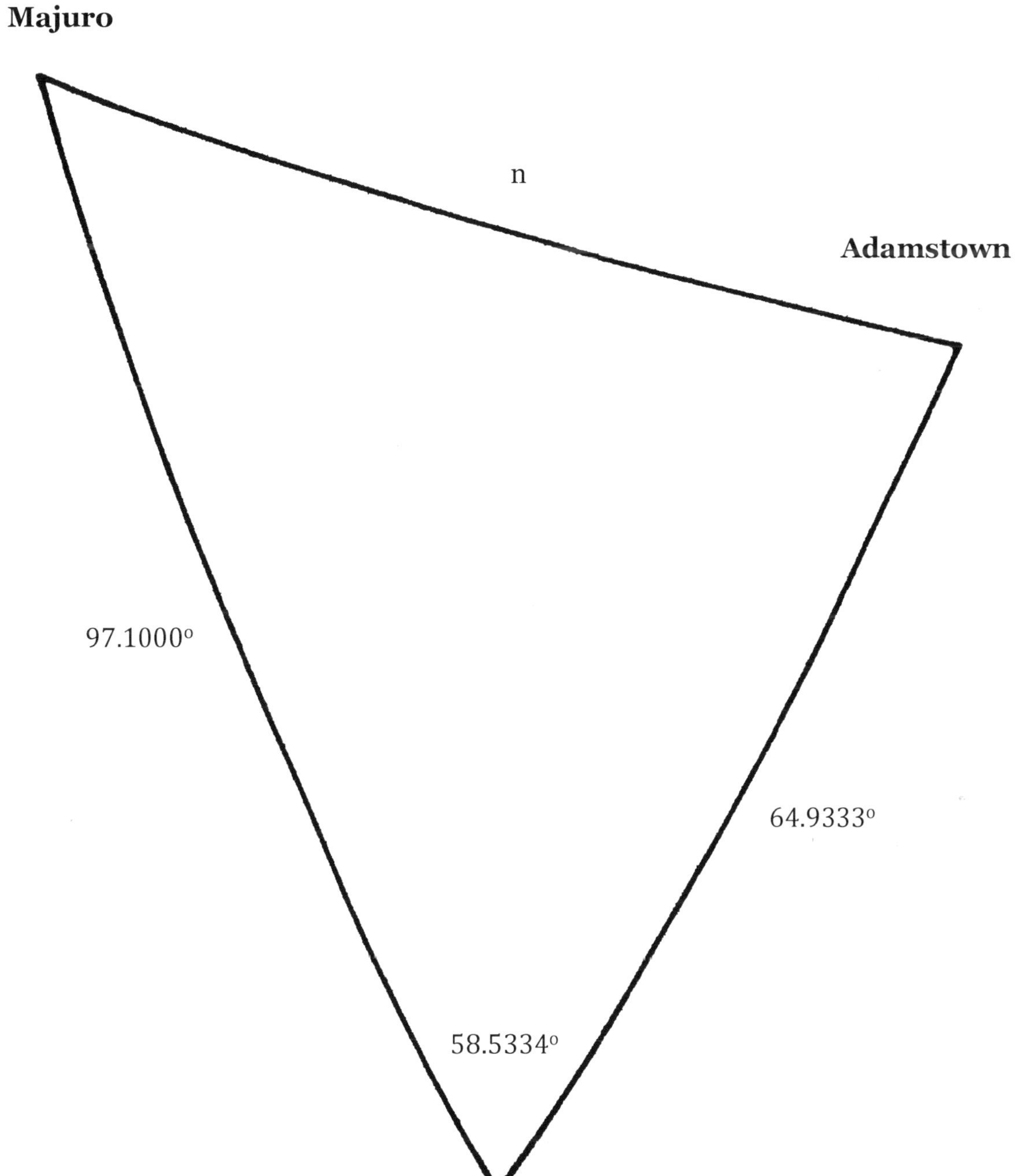
Majuro
n
Adamstown
97.1000°
64.9333°
58.5334°
South Pole

Example 3

Ngerulmud, Palau is located at $7^{o}30' \text{ N} = \left(7 + {}^{30}/_{60}\right)^{o} \text{ N} \approx 7.500^{o} \text{ N}$, $134^{o}37' \text{ E} = \left(134 + {}^{37}/_{60}\right)^{o} \text{ E} \approx 134.6167^{o} \text{ E}$.

Hagatna, Guam is located at $13^{o}28' \text{ N} = \left(13 + {}^{28}/_{60}\right)^{o} \text{ N} \approx 13.4667^{o} \text{ N}$, $144^{o}44' \text{ E} = \left(144 + {}^{44}/_{60}\right)^{o} \text{ E} \approx 144.7333^{o} \text{ E}$.

Two sides of the spherical triangle are $90^{o} - 7.5000^{o} = 82.5000^{o}$ and $90^{o} - 13.4667^{o} = 76.5333^{o}$, respectively. The angle at the North Pole is $144.7333^{o} - 134.6167^{o} = 10.1166^{o}$.

See the illustration on the next page.

It follows that

$\cos n = \cos(82.5000^{o}) \cos(76.5333^{o}) + \sin(82.5000^{o}) \ \sin(76.5333^{o}) \cos(10.1166^{o})$

$\cos n \approx 0.9795916411 \ \Rightarrow \ n \approx \cos^{-1}(0.9795916411) \ \Rightarrow \ n \approx 11.59532841^{o}$.

Therefore, the approximate distance between the two capital cities is

$(11.59532841^{o})\left({}^{\pi}/_{180^{o}}\right)(3959 \text{ miles})$

$\approx (0.2023766586)(3959 \text{ miles})$

≈ 801 miles.

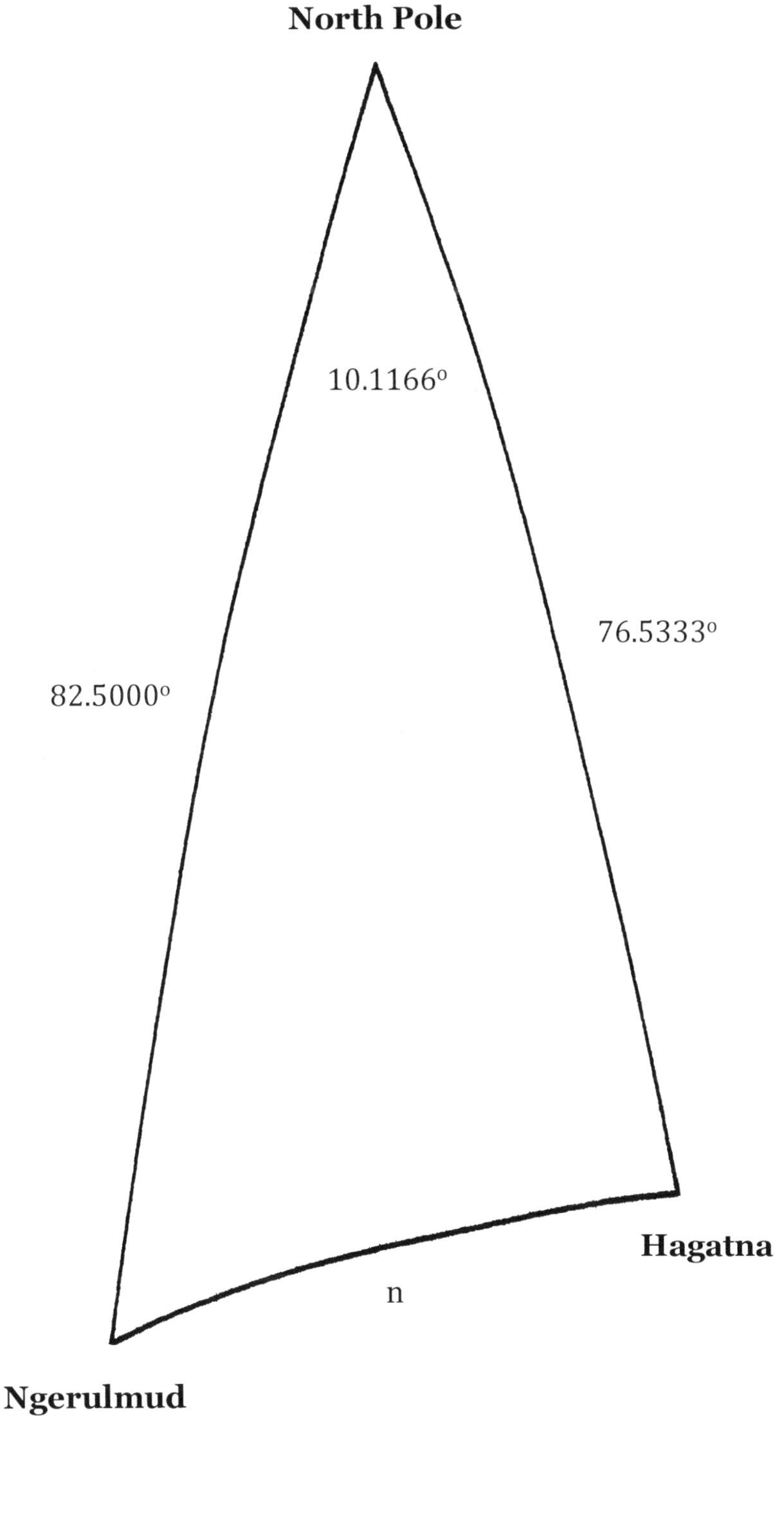
North Pole
10.1166°
76.5333°
82.5000°
Hagatna
n
Ngerulmud

Chapter 3 Problems:

Approximate the distance between the following capitals in Australia & Oceania.

1. Canberra, Australia (35°16′ S 149°8′ E)
   to Wellington, New Zealand (41°18′ S 174°47′ E)

2. West Island, Cocos (Keeling) Islands (12°10′ S 96°50′ E)
   to Pago Pago, American Samoa (14°16′ S 170°42′ W)

3. Avarua, Cook Islands (21°12′S 159°46′W)
   to Papeete, French Polynesia (17°32′S 149°34′W)

4. Tarawa, Kiribati (1°21′ N 173°2′ E)
   to Suva, Fiji (18°8′ S 178°25′ E)

5. Noumea, New Caledonia (22°16′ S 166°27′ E)
   to Mata-Utu, Wallis and Futuna (13°57′ S 171°56′ W)

6. Kingston, Norfolk Island (29°3′ S 167°58′ E)
   to Apia, Samoa (13°49′ S 171°46′ W)

7. Saipan, Northern Mariana Islands (15°12′ N 145°45′ E)
   to Palikir, Federated States of Micronesia (6°55′ N 158°9′ E)

8. Honiara, Solomon Islands (9°26′ S 159°57′ E)
   to Port-Vila, Vanuatu (17°44′ S 168°19′ E)

9. Funafuti, Tuvalu (8°31′ S 179°13′ E)
   to Alofi, Niue (19°1′ S 169°55′ W)

Chapter 3
Problem 1
Solution

Canberra, Australia is located at $35^{o}16' \text{ S} = \left(35 + {}^{16}/_{60}\right)^{o} \text{ S} \approx 35.2667^{o} \text{ S}$,

$149^{o}8' \text{ E} = \left(149 + {}^{8}/_{60}\right)^{o} \text{ E} \approx 149.1333^{o} \text{ E}$.

Wellington, New Zealand is located at $41^{o}18' \text{ S} = \left(41 + {}^{18}/_{60}\right)^{o} \text{ S} = 41.3000^{o} \text{ S}$,

$174^{o}47' \text{ E} = \left(174 + {}^{47}/_{60}\right)^{o} \text{ E} \approx 174.7833^{o} \text{ E}$.

Two sides of the spherical triangle are $90^{o} - 35.2667^{o} = 54.7333^{o}$ and $90^{o} - 41.3000^{o} = 48.7000^{o}$, respectively. The angle at the South Pole is $174.7833^{o} - 149.1333^{o} = 25.6500^{o}$.

See the illustration on the next page.

It follows that

$\cos n = \cos(54.7333^{o}) \cos(48.7000^{o}) + \sin(54.7333^{o}) \sin(48.7000^{o}) \cos(25.6500^{o})$

$\cos n \approx 0.934014815472 \Rightarrow n \approx \cos^{-1}(0.934014815472) \Rightarrow n \approx 20.9304399815^{o}$.

Therefore, the approximate distance between the two capital cities is

$(20.9304399815^{o})\left({}^{\pi}/_{180^{o}}\right)(3959 \text{ miles})$

$\approx (0.365305091568)(3959 \text{ miles})$

$\approx 1{,}446 \text{ miles}$.

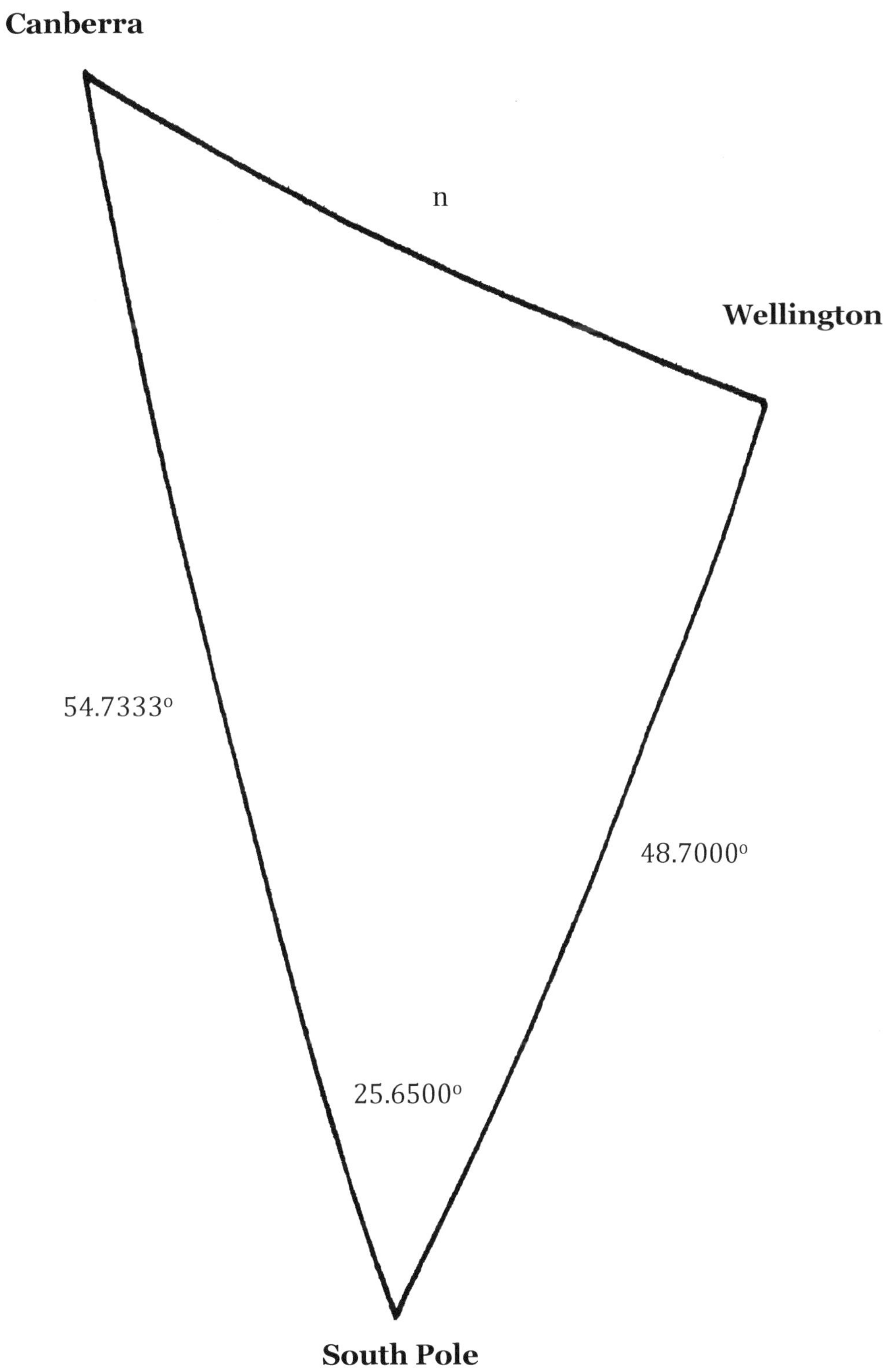
Canberra
n
Wellington
54.7333°
48.7000°
25.6500°
South Pole

Chapter 3
Problem 2
Solution

West Island, Cocos (Keeling) Islands is located at

$12^o10' \text{ S} = \left(12 + {}^{10}/_{60}\right)^o \text{ S} \approx 12.1667^o \text{ S}$, $96^o50' \text{ E} = \left(96 + {}^{50}/_{60}\right)^o \text{ E} \approx 96.8333^o \text{ E}$.

Pago Pago, American Samoa is located at $14^o16' \text{ S} = \left(14 + {}^{16}/_{60}\right)^o \text{ S} \approx 14.2667^o \text{ S}$, $170^o42' \text{ W} = \left(170 + {}^{42}/_{60}\right)^o \text{ W} = 170.7000^o \text{ W}$.

Two sides of the spherical triangle are $90^o - 12.1667^o = 77.8333^o$ and $90^o - 14.2667^o = 75.7333^o$, respectively. The angle at the South Pole is $(180^o - 96.8333^o) + (180^o - 170.7000^o) = 83.1667^o + 9.3000^o = 92.4667^o$.

See the illustration on the next page.

It follows that

$\cos n = \cos(77.8333^o)\cos(75.7333^o) + \sin(77.8333^o)\ \sin(75.7333^o)\cos(92.4667^o)$

$\cos n \approx 0.011163502794 \Rightarrow n \approx \cos^{-1}(0.011163502794) \Rightarrow n \approx 89.3603651192^o$.

Therefore, the approximate distance between the two capital cities is

$(89.3603651192^o)\left({}^{\pi}/_{180^o}\right)(3959 \text{ miles})$

$\approx (1.55963259212)(3959 \text{ miles})$

$\approx 6{,}175$ miles.

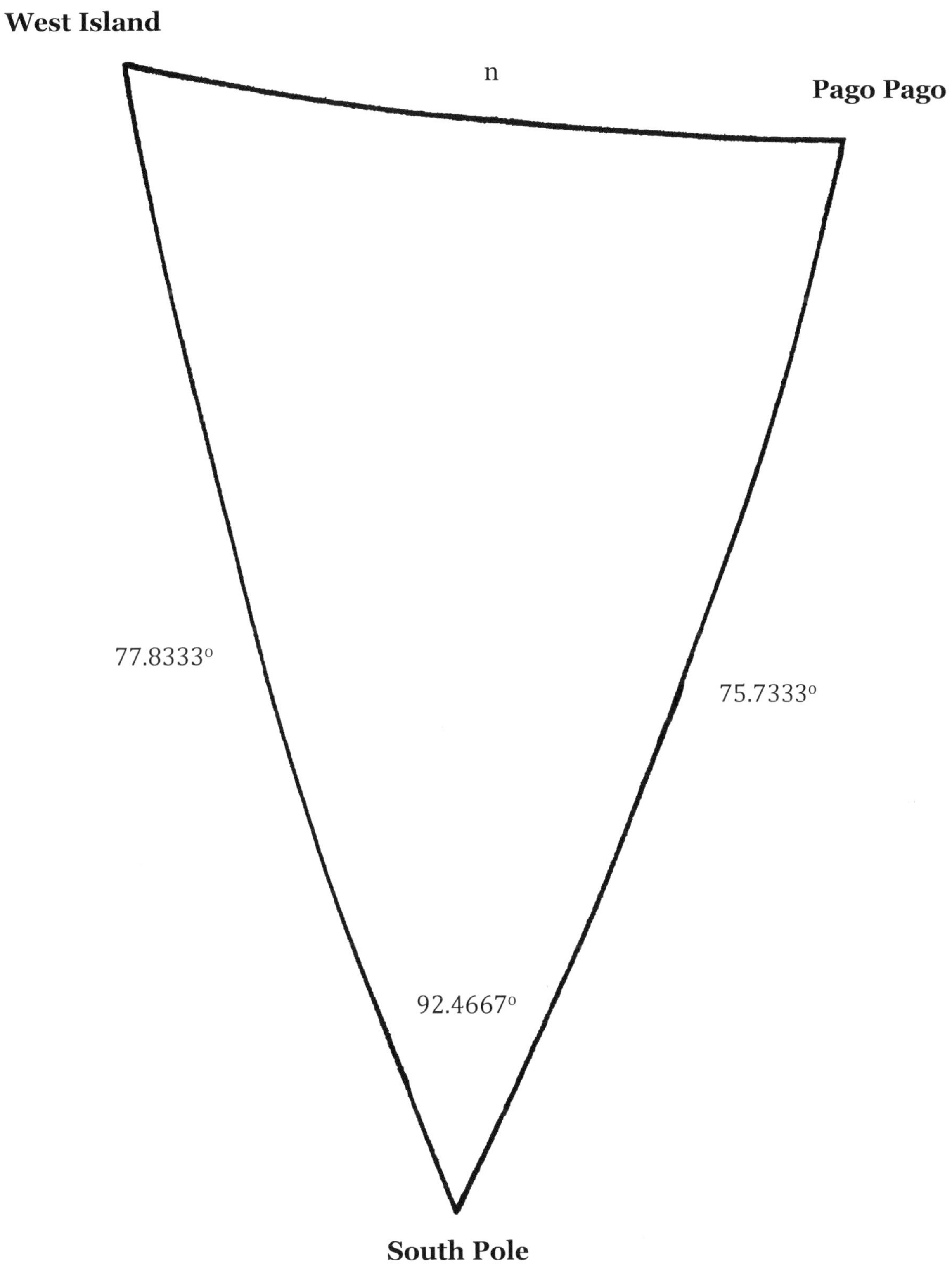
West Island
n
Pago Pago
77.8333°
75.7333°
92.4667°
South Pole

Chapter 3
Problem 3
Solution

Avarua, Cook Islands is located at $21^o12'S = \left(21 + {}^{12}/_{60}\right)^o S = 21.2000^o S$,

$159^o46'W = \left(159 + {}^{46}/_{60}\right)^o W \approx 159.7667^o W$.

Papeete, French Polynesia is located at $17^o32'S = \left(17 + {}^{32}/_{60}\right)^o S \approx 17.5333^o S$,

$149^o34'W = \left(149 + {}^{34}/_{60}\right)^o W \approx 149.5667^o W$.

Two sides of the spherical triangle are $90^o - 21.2000^o = 68.8000^o$ and $90^o - 17.5333^o = 72.4667^o$, respectively. The angle at the South Pole is $159.7667^o - 149.5667^o = 10.2000^o$.

See the illustration on the next page.

It follows that

$\cos n = \cos(68.8000^o)\cos(72.4667^o) + \sin(68.8000^o)\sin(72.4667^o)\cos(10.2000^o)$

$\cos n \approx 0.983902694238 \Rightarrow n \approx \cos^{-1}(0.983902694238) \Rightarrow n \approx 10.2943406365^o$.

Therefore, the approximate distance between the two capital cities is

$(10.2943406365^o)\left({}^{\pi}/_{180^o}\right)(3959 \text{ miles})$

$\approx (0.179670138428)(3959 \text{ miles})$

≈ 711 miles.

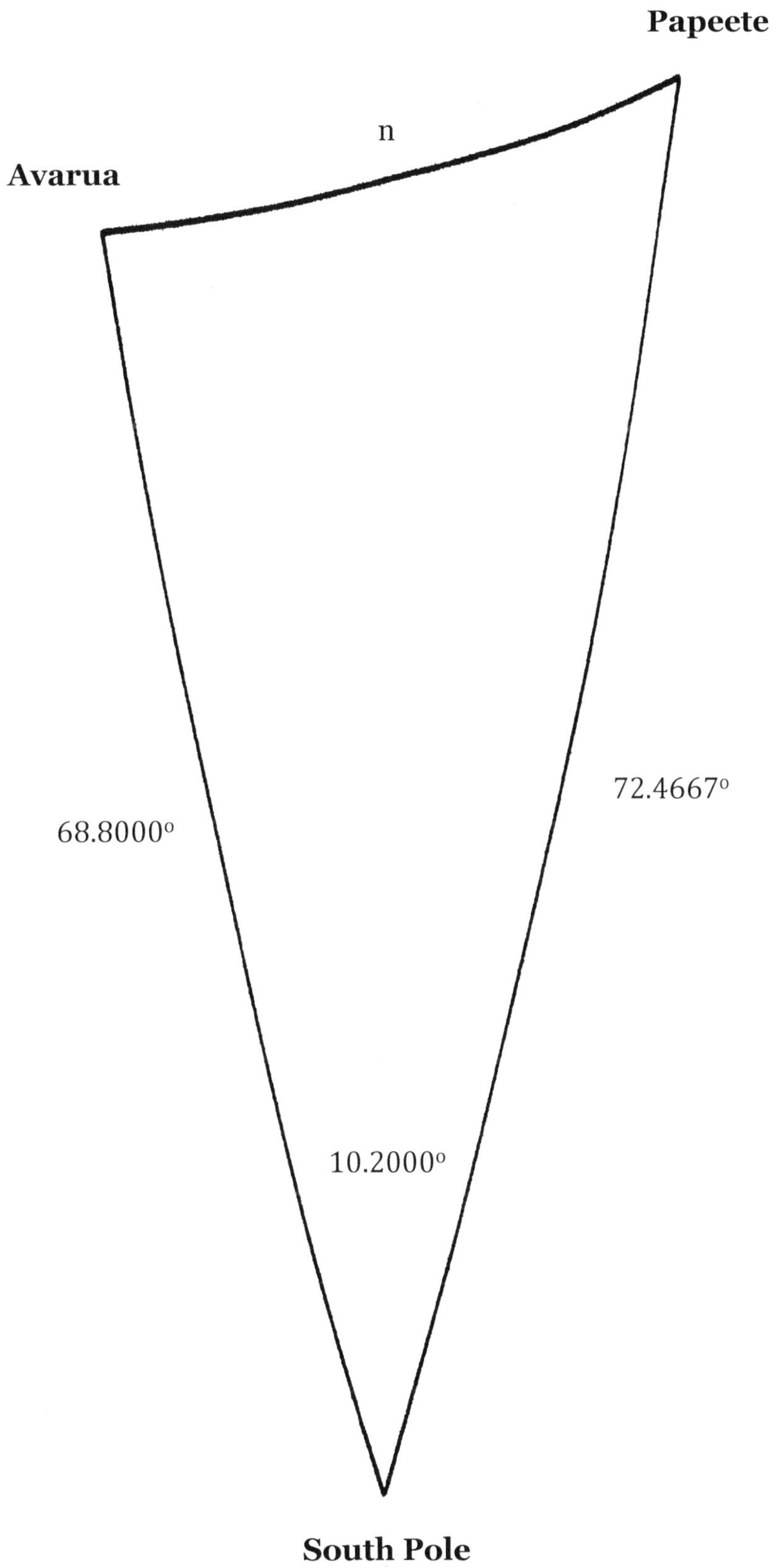

Papeete
n
Avarua
72.4667°
68.8000°
10.2000°
South Pole

Chapter 3
Problem 4
Solution

Tarawa, Kiribati is located at $1^o21'\text{ N} = \left(1 + {}^{21}/_{60}\right)^o \text{ N} = 1.3500^o \text{ N}$,

$173^o2'\text{ E} = \left(173 + {}^{2}/_{60}\right)^o \text{ E} \approx 173.0333^o \text{ E}$.

Suva, Fiji is located at $18^o8'\text{ S} = \left(18 + {}^{8}/_{60}\right)^o \text{ S} \approx 18.1333^o \text{ S}$,

$178^o25'\text{ E} = \left(178 + {}^{25}/_{60}\right)^o \text{ E} \approx 178.4167^o \text{ E}$.

Two sides of the spherical triangle are $90^o + 1.3500^o = 91.3500^o$ and $90^o - 18.1333^o = 71.8667^o$, respectively. The angle at the South Pole is $178.4167^o - 173.0333^o = 5.3834^o$.

See the illustration on the next page.

It follows that

$$\cos n = \cos(91.3500^o)\cos(71.8667^o) + \sin(91.3500^o)\ \sin(71.8667^o)\cos(5.3834^o)$$

$$\cos n \approx 0.938548160479 \ \Rightarrow\ n \approx \cos^{-1}(0.938548160479) \ \Rightarrow\ n \approx 20.1908489402^o.$$

Therefore, the approximate distance between the two capital cities is

$$(20.1908489402^o)\left({}^{\pi}/_{180^o}\right)(3959 \text{ miles})$$

$$\approx (0.352396792779)(3959 \text{ miles})$$

$$\approx\ 1{,}395\ \text{ miles.}$$

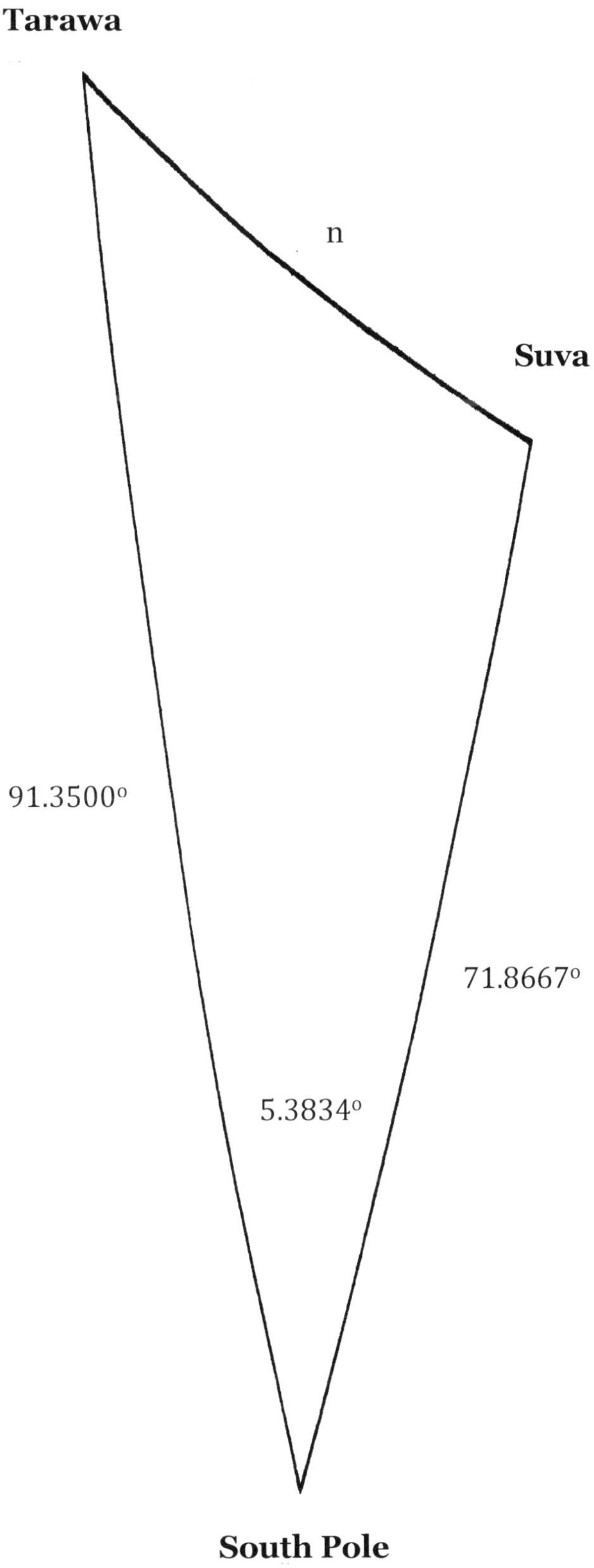
Tarawa
n
Suva
91.3500°
71.8667°
5.3834°
South Pole

Chapter 3
Problem 5
Solution

Noumea, New Caledonia is located at $22^o16'\ \mathrm{S} = \left(22 + {}^{16}/_{60}\right)^o\ \mathrm{S} \approx 22.2667^o\ \mathrm{S}$,

$166^o27'\ \mathrm{E} = \left(166 + {}^{27}/_{60}\right)^o\ \mathrm{E} = 166.4500^o\ \mathrm{E}$.

Mata-Utu, Wallis and Futuna is located at

$13^o57'\ \mathrm{S} = \left(13 + {}^{57}/_{60}\right)^o\ \mathrm{S} = 13.9500^o\ \mathrm{S}$,

$171^o56'\ \mathrm{W} = \left(171 + {}^{56}/_{60}\right)^o\ \mathrm{W} \approx 171.9333^o\ \mathrm{W}$.

Two sides of the spherical triangle are $90^o - 22.2667^o = 67.7333^o$ and $90^o - 13.9500^o = 76.0500^o$, respectively. The angle at the South Pole is $(180^o - 166.4500^o) + (180^o - 171.9333^o) = 13.5500^o + 8.0667^o = 21.6167^o$.

See the illustration on the next page.

It follows that

$\cos n = \cos(67.7333^o)\cos(76.0500^o) + \sin(67.7333^o)\ \sin(76.0500^o)\cos(21.6167^o)$

$\cos n \approx 0.926317009304\ \Rightarrow\ n \approx \cos^{-1}(0.926317009304)\ \Rightarrow\ n \approx 22.1322058891^o$.

Therefore, the approximate distance between the two capital cities is

$(22.1322058891^o)\left({}^{\pi}/_{180^o}\right)(3959\ \text{miles})$

$\approx (0.386279863495)(3959\ \text{miles})$

$\approx\ 1{,}529\ \text{miles}$.

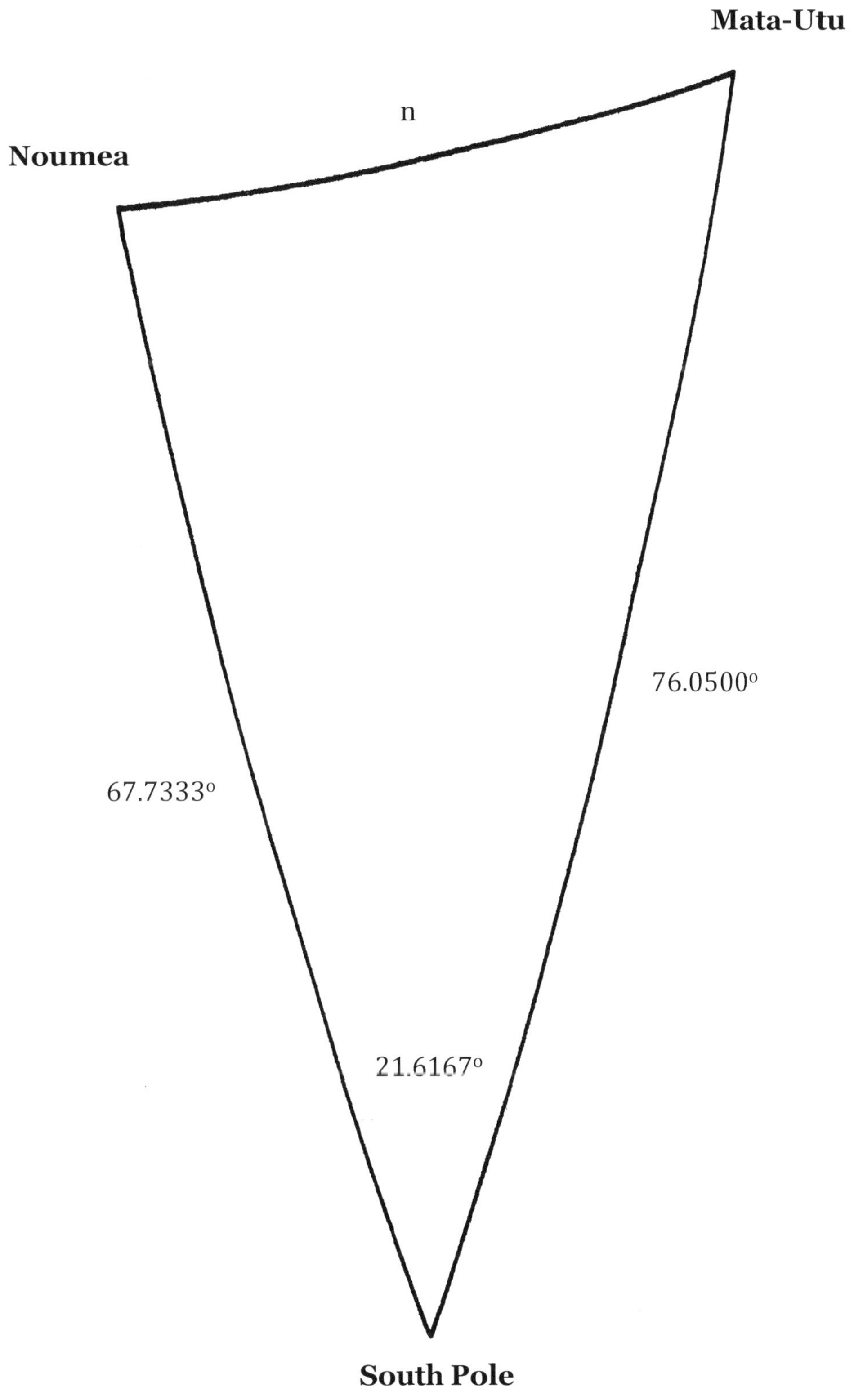
Mata-Utu
n
Noumea
76.0500°
67.7333°
21.6167°
South Pole

Chapter 3
Problem 6
Solution

Kingston, Norfolk Island is located at $29^o 3' \text{ S} = \left(29 + {}^{3}/_{60}\right)^o \text{ S} = 29.0500^o \text{ S}$,

$167^o 58' \text{ E} = \left(167 + {}^{58}/_{60}\right)^o \text{ E} \approx 167.9667^o \text{ E}$.

Apia, Samoa is located at $13^o 49' \text{ S} = \left(13 + {}^{49}/_{60}\right)^o \text{ S} \approx 13.8167^o \text{ S}$,

$171^o 46' \text{ W} = \left(171 + {}^{46}/_{60}\right)^o \text{ W} \approx 171.7667^o \text{ W}$.

Two sides of the spherical triangle are $90^o - 29.0500^o = 60.9500^o$ and $90^o - 13.8167^o = 76.1833^o$, respectively. The angle at the South Pole is $(180^o - 167.9667^o) + (180^o - 171.7667^o) = 12.0333^o + 8.2333^o \approx 20.2666^o$.

See the illustration on the next page.

It follows that

$\cos n = \cos(60.9500^o)\cos(76.1833^o) + \sin(60.9500^o)\ \sin(76.1833^o)\cos(20.2666^o)$

$\cos n \approx 0.912309340753 \Rightarrow n \approx \cos^{-1}(0.912309340753) \Rightarrow n \approx 24.1735380962^o$.

Therefore, the approximate distance between the two capital cities is

$(24.1735380962^o)\left({}^{\pi}/_{180^o}\right)(3959 \text{ miles})$

$\approx (0.421907831635)(3959 \text{ miles})$

$\approx$ 1,670 miles.

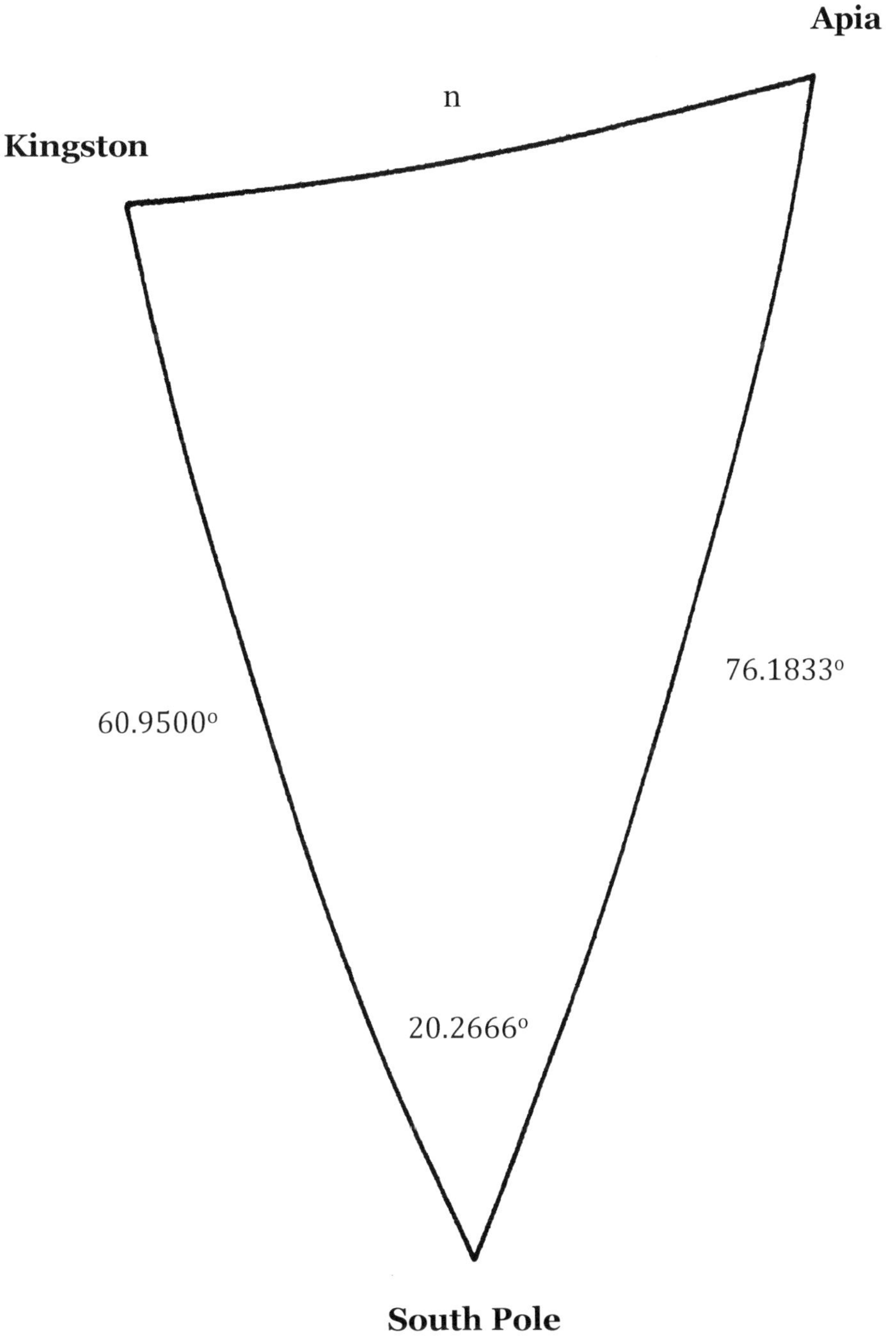
Apia
n
Kingston
76.1833°
60.9500°
20.2666°
South Pole

Chapter 3
Problem 7
Solution

Saipan, Northern Mariana Islands is located at

$15^o12'\text{ N} = \left(15 + {}^{12}/_{60}\right)^o\text{ N} = 15.2000^o\text{ N}$,

$145^o45'\text{ E} = \left(145 + {}^{45}/_{60}\right)^o\text{ E} = 145.7500^o\text{ E}$.

Palikir, Federated States of Micronesia is located at

$6^o55'\text{ N} = \left(6 + {}^{55}/_{60}\right)^o\text{ N} \approx 6.9167^o\text{ N}$, $158^o9'\text{ E} = \left(158 + {}^{9}/_{60}\right)^o\text{ E} = 158.1500^o\text{ E}$.

Two sides of the spherical triangle are $90^o - 15.2000^o = 74.8000^o$ and $90^o - 6.9167^o = 83.0833^o$, respectively. The angle at the North Pole is $158.1500^o - 145.7500^o = 12.4000^o$.

See the illustration on the next page.

It follows that

$\cos n = \cos(74.8000^o)\cos(83.0833^o) + \sin(74.8000^o)\sin(83.0833^o)\cos(12.4000^o)$

$\cos n \approx 0.967220019656 \Rightarrow n \approx \cos^{-1}(0.967220019656) \Rightarrow n \approx 14.7107737574^o$.

Therefore, the approximate distance between the two capital cities is

$(14.7107737574^o)\left({}^{\pi}/_{180^o}\right)(3959\text{ miles})$

$\approx (0.256751437583)(3959\text{ miles})$

$\approx 1{,}016$ miles.

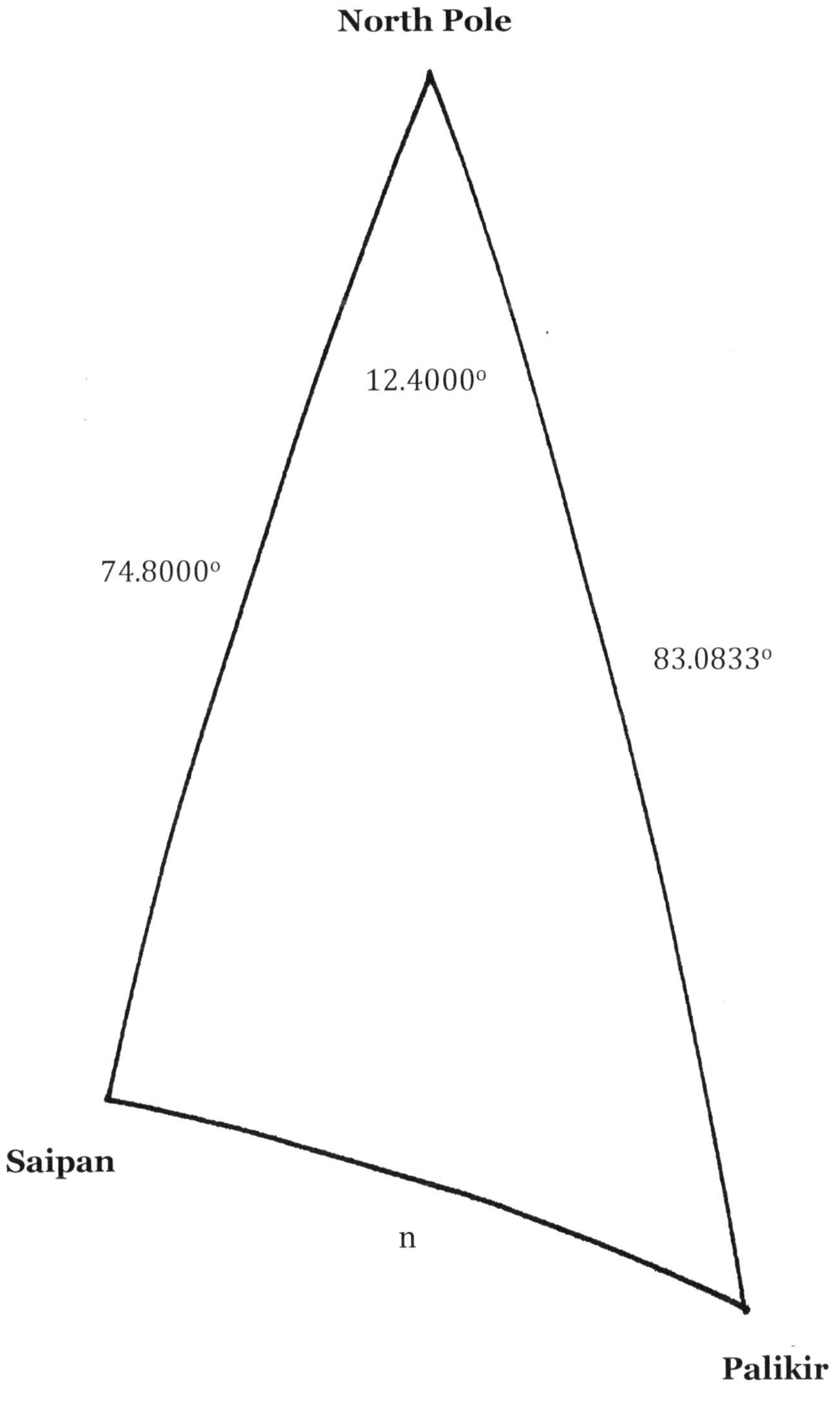
North Pole
12.4000°
74.8000°
83.0833°
Saipan
n
Palikir

Chapter 3
Problem 8
Solution

Honiara, Solomon Islands is located at $9^o26' \text{ S} = \left(9 + {}^{26}/_{60}\right)^o \text{ S} \approx 9.4333^o \text{ S}$,

$159^o57' \text{ E} = \left(159 + {}^{57}/_{60}\right)^o \text{ E} = 159.9500^o \text{ E}$.

Port-Vila, Vanuatu is located at $17^o44' \text{ S} = \left(17 + {}^{44}/_{60}\right)^o \text{ S} \approx 17.7333^o \text{ S}$,

$168^o19' \text{ E} = \left(168 + {}^{19}/_{60}\right)^o \text{ E} \approx 168.3167^o \text{ E}$.

Two sides of the spherical triangle are $90^o - 9.4333^o = 80.5667^o$ and $90^o - 17.7333^o = 72.2667^o$, respectively. The angle at the South Pole is $168.3167^o - 159.9500^o = 8.3667^o$.

See the illustration on the next page.

It follows that

$\cos n = \cos(80.5667^o)\cos(72.2667^o) + \sin(80.5667^o)\sin(72.2667^o)\cos(8.3667^o)$

$\cos n \approx 0.979525645596 \Rightarrow n \approx \cos^{-1}(0.979525645596) \Rightarrow n \approx 11.614125839^o$.

Therefore, the approximate distance between the two capital cities is

$(11.614125839^o)\left({}^{\pi}/_{180^o}\right)(3959 \text{ miles})$

$\approx (0.202704735632)(3959 \text{ miles})$

≈ 803 miles.

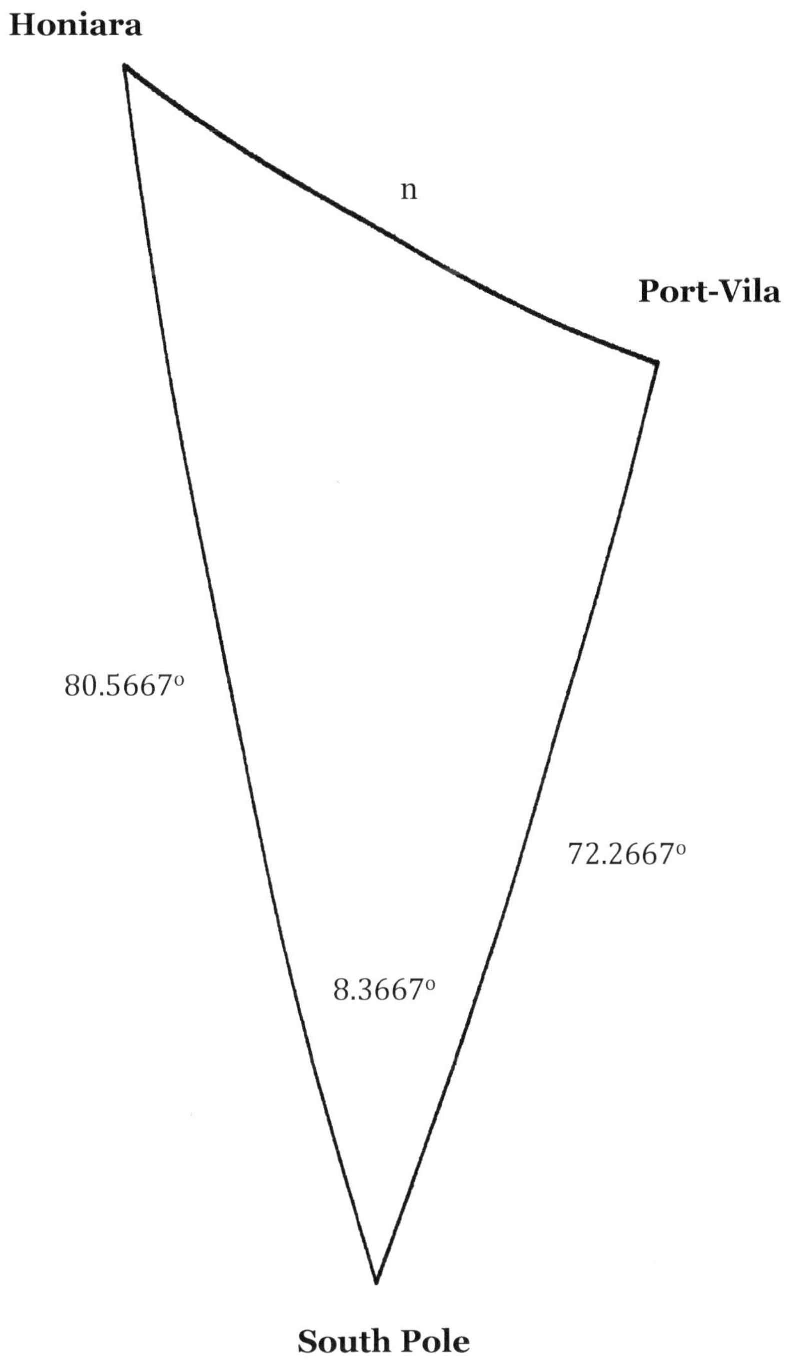
Honiara
n
Port-Vila
80.5667°
72.2667°
8.3667°
South Pole

Chapter 3
Problem 9
Solution

Funafuti, Tuvalu is located at $8^o31' \text{ S} = \left(8 + {}^{31}/_{60}\right)^o \text{ S} \approx 8.5167^o \text{ S}$,

$179^o13' \text{ E} = \left(179 + {}^{13}/_{60}\right)^o \text{ E} \approx 179.2167^o \text{ E}$.

Alofi, Niue is located at $19^o1' \text{ S} = \left(19 + {}^{1}/_{60}\right)^o \text{ S} \approx 19.0167^o \text{ S}$,

$169^o55' \text{ W} = \left(169 + {}^{55}/_{60}\right)^o \text{ W} \approx 169.9167^o \text{ W}$.

Two sides of the spherical triangle are $90^o - 8.5167^o = 81.4833^o$ and $90^o - 19.0167^o = 70.9833^o$, respectively. The angle at the South Pole is $(180^o - 179.2167^o) + (180^o - 169.9167^o) = 0.7833^o + 10.0833^o = 10.8666^o$.

See the illustration on the next page.

It follows that

$\cos n = \cos(81.4833^o)\cos(70.9833^o) + \sin(81.4833^o)\ \sin(70.9833^o)\cos(10.8666^o)$

$\cos n \approx 0.9664892463 \ \Rightarrow\ n \approx \cos^{-1}(0.9664892463) \ \Rightarrow\ n \approx 14.8747626053^o$.

Therefore, the approximate distance between the two capital cities is

$(14.8747626053^o)\left({}^{\pi}/_{180^o}\right)(3959 \text{ miles})$

$\approx (0.259613582914)(3959 \text{ miles})$

$\approx\ 1{,}028\ \text{ miles.}$

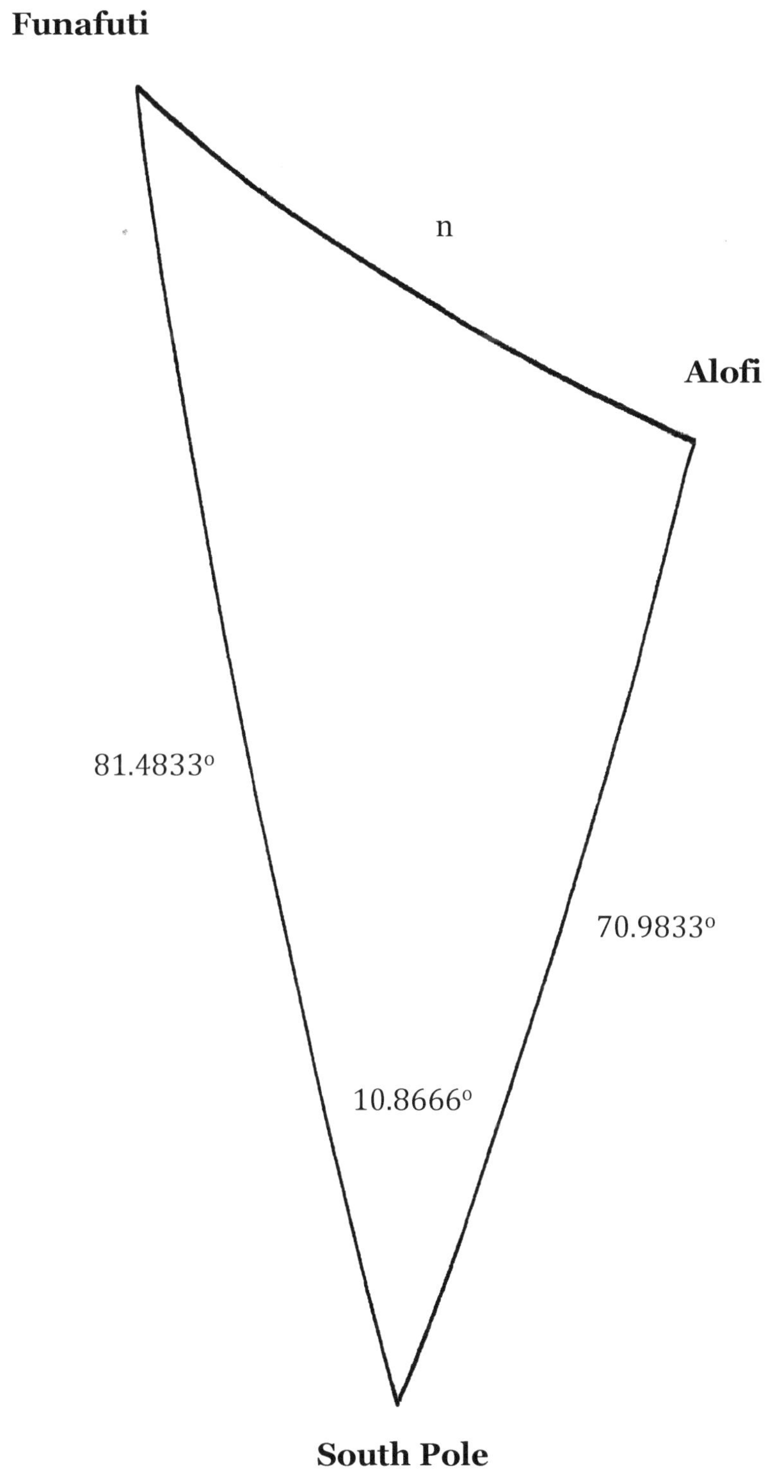
Funafuti
n
Alofi
81.4833°
70.9833°
10.8666°
South Pole

Chapter 4

Central America & the Caribbean

Example 1

George Town, Cayman Islands is located at

$19^o18' \text{ N} = \left(19 + {}^{18}/_{60}\right)^o \text{ N} = 19.3000^o \text{ N}$,

$81^o23' \text{ W} = \left(81 + {}^{23}/_{60}\right)^o \text{ W} \approx 81.3833^o \text{ W}$.

Port-au-Prince, Haiti is located at $18^o32' \text{ N} = \left(18 + {}^{32}/_{60}\right)^o \text{ N} \approx 18.5333^o \text{ N}$,

$72^o20' \text{ W} = \left(72 + {}^{20}/_{60}\right)^o \text{ W} \approx 72.3333^o \text{ W}$.

Two sides of the spherical triangle are $90^o - 19.3000^o = 70.7000^o$ and $90^o - 18.5333^o = 71.4667^o$, respectively. The angle at the North Pole is $81.3833^o - 72.3333^o = 9.0500^o$.

See the illustration on the next page.

It follows that

$\cos n = \cos(70.7000^o)\cos(71.4667^o) + \sin(70.7000^o)\ \sin(71.4667^o)\cos(9.0500^o)$

$\cos n \approx 0.988770827813 \ \Rightarrow\ n \approx \cos^{-1}(0.988770827813) \ \Rightarrow\ n \approx 8.59446282543^o$.

Therefore, the approximate distance between the two capital cities is

$(8.59446282543^o)\left({}^{\pi}/_{180^o}\right)(3959 \text{ miles})$

$\approx (0.150001673744)(3959 \text{ miles})$

≈ 594 miles.

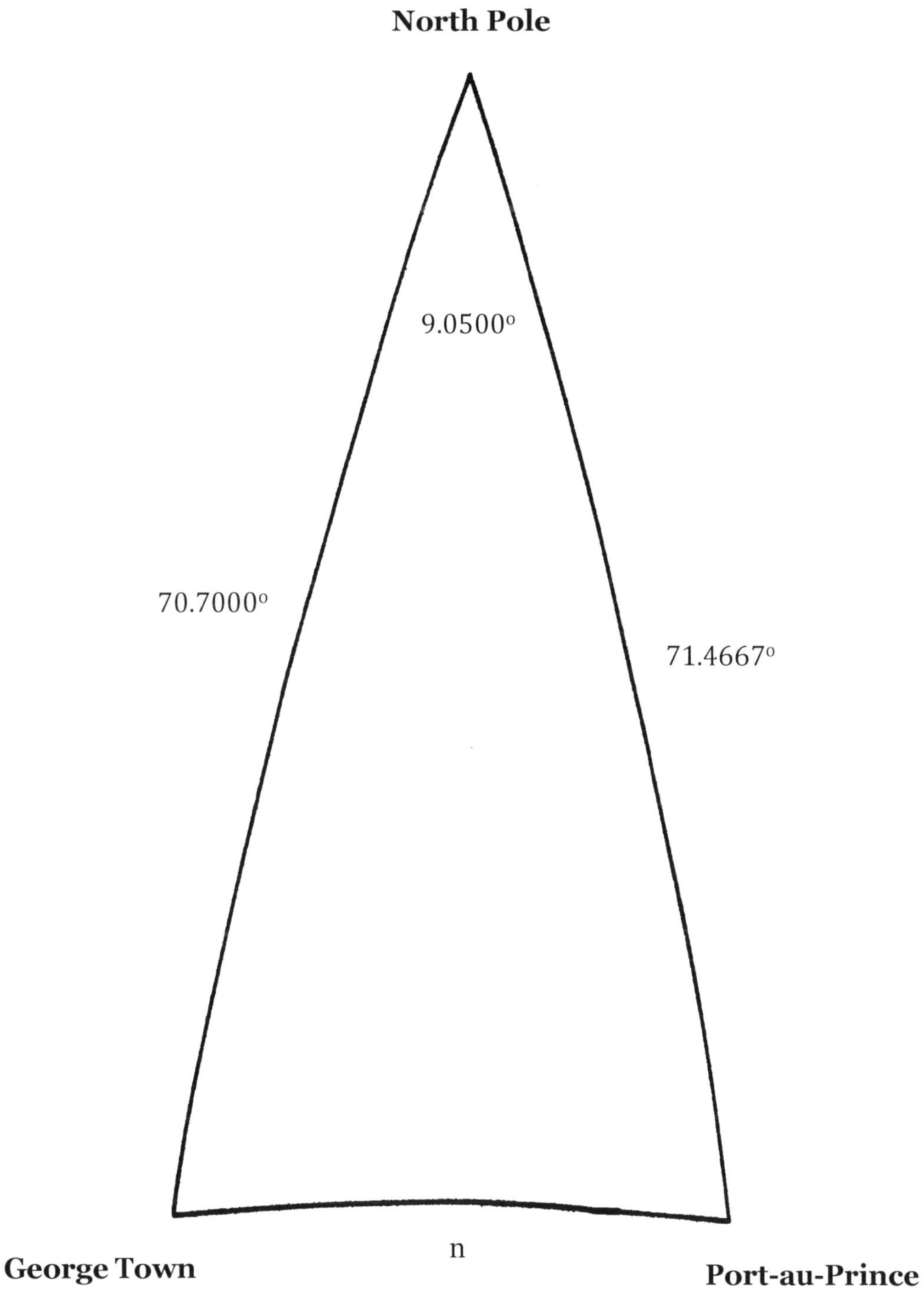
North Pole
9.0500°
70.7000°
71.4667°
George Town
n
Port-au-Prince

Example 2

Guatemala City, Guatemala is located at $14^o37'$ N $= \left(14 + {}^{37}/_{60}\right)^o$ N $\approx 14.6167^o$ N ,

$90^o31'$ W $= \left(90 + {}^{31}/_{60}\right)^o$ W $\approx 90.5167^o$ W .

Managua, Nicaragua is located at $12^o8'$ N $= \left(12 + {}^{8}/_{60}\right)^o$ N $\approx 12.1333^o$ N ,

$86^o15'$ W $= \left(86 + {}^{15}/_{60}\right)^o$ W $= 86.2500^o$ W .

Two sides of the spherical triangle are $90^o - 14.6167^o = 75.3833^o$ and $90^o - 12.1333^o = 77.8667^o$, respectively. The angle at the North Pole is $90.5167^o - 86.2500^o = 4.2667^o$.

See the illustration on the next page.

It follows that

$\cos n = \cos(75.3833^o)\cos(77.8667^o) + \sin(75.3833^o)\ \sin(77.8667^o)\cos(4.2667^o)$

$\cos n \approx 0.996438963218 \Rightarrow n \approx \cos^{-1}(0.996438963218) \Rightarrow n \approx 4.83676316508^o$.

Therefore, the approximate distance between the two capital cities is

$(4.83676316508^o)\left({}^{\pi}/_{180^o}\right)(3959 \text{ miles})$

$\approx (0.08441744237)(3959 \text{ miles})$

≈ 334 miles.

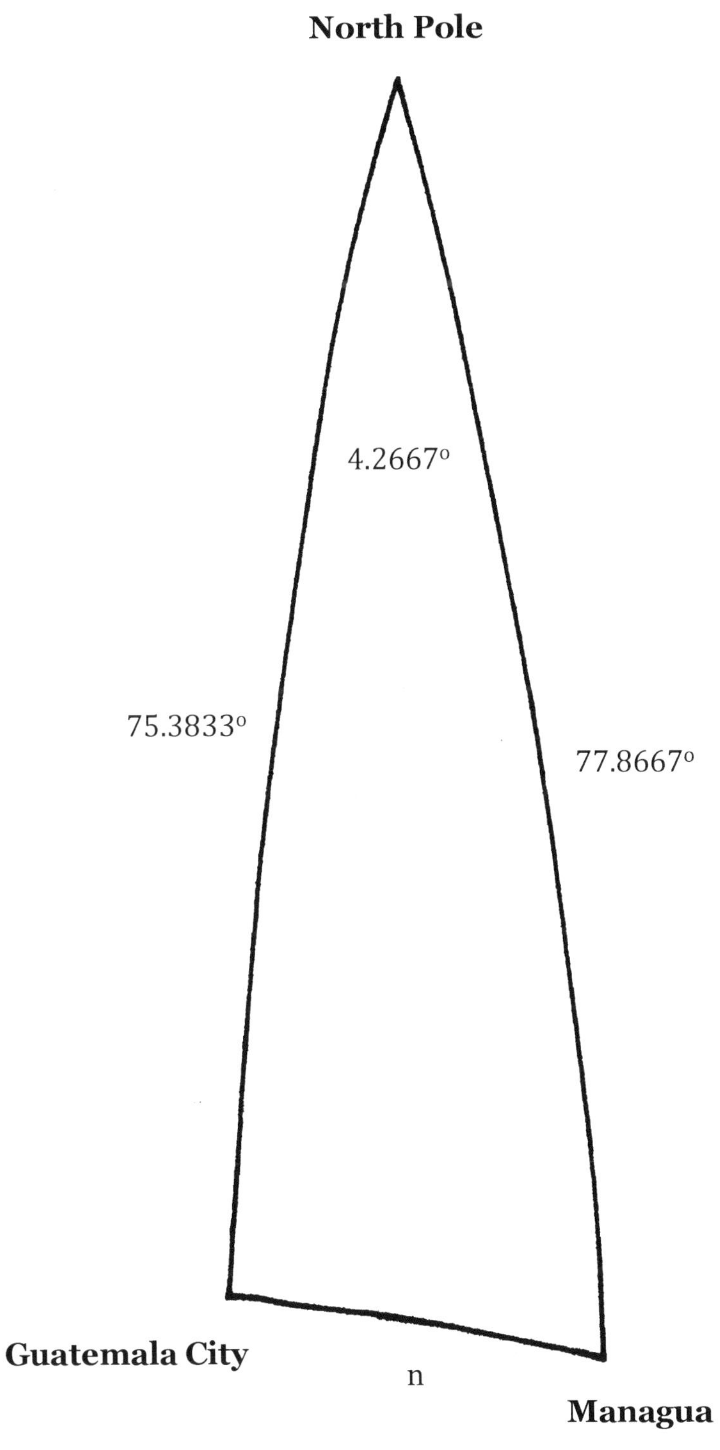
North Pole
4.2667°
75.3833°
77.8667°
Guatemala City
n
Managua

Chapter 4 Problems:

Approximate the distance between capitals in Central American and the Caribbean.

1. The Valley, Anguilla (18°13′ N 63°3′ W)
   to Castries, Saint Lucia (14°0′ N 61°0′ W)

2. Oranjestad, Aruba (12°31′ N 70°2′ W)
   to Port of Spain, Trinidad and Tobago (10°39′N 61°31′ W)

3. Belmopan, Belize (17°15′ N 88°46′ W)
   to Kingston, Jamaica (18°0′ N 76°48′ W)

4. Havana, Cuba (23°7′ N 82°21′ W)
   to Bridgetown, Barbados (13°6′ N 59°37′ W)

5. Willemstad, Curacao (12°6′ N 68°55′ W)
   to Saint Georges, Grenada (12°3′ N 61°45′ W)

6. San Salvador, El Salvador (13°42′ N 89°12′ W)
   to Road Town, British Virgin Islands (18°25′ N 64°37′ W)

7. Tegucigalpa, Honduras (14°6′ N 87°13′ W)
   to San Jose, Costa Rica (9°56′ N 84°5′ W)

8. Plymouth, Montserrat (16°42′ N 62°13′ W)
   to Roseau, Dominica (15°18′ N 61°24′ W)

9. Panama City, Panama (8°58′ N 79°32′ W)
   to Nassau, The Bahamas (25°5′ N 77°21′ W)

10. San Juan, Puerto Rico (18°28′ N 66°7′ W)
    to Nassau, The Bahamas (25°5′ N 77°21′ W)

11. Marigot, Saint Martin (18°4′ N 63°5′ W)
    to Gustavia, Saint Barthelemy (17°53′ N 62°51′ W)

12. Philipsburg, Sint Maarten (18°1′ N 63°2′ W)
    to Saint John's, Antigua and Barbuda (17°7′ N 61°51′ W)

13. Grand Turk, Turks and Caicos Islands (21°28′ N 71°8′ W)
    to Santo Domingo, Dominican Republic (18°28′ N 69°54′ W)

14. Charlotte Amalie, U.S. Virgin Islands (18°21′ N 64°56′ W)
to Basseterre, Saint Kitts and Nevis (17°18′ N 62°43′ W)

Chapter 4
Problem 1
Solution

The Valley, Anguilla is located at $18^{o}13' \text{ N} = \left(18 + {}^{13}/_{60}\right)^{o} \text{ N} \approx 18.2167^{o} \text{ N}$,

$63^{o}3' \text{ W} = \left(63 + {}^{3}/_{60}\right)^{o} \text{ W} = 63.0500^{o} \text{ W}$.

Castries, Saint Lucia is located at $14^{o}0' \text{ N} = 14^{o} \text{ N}$, $61^{o}0' \text{ W} = 61^{o} \text{ W}$.

Two sides of the spherical triangle are $90^{o} - 18.2167^{o} = 71.7833^{o}$ and $90^{o} - 14^{o} = 76^{o}$, respectively. The angle at the North Pole is $63.0500^{o} - 61^{o} = 2.0500^{o}$.

See the illustration on the next page.

It follows that

$\cos \text{n} = \cos(71.7833^{o}) \cos(76^{o}) + \sin(71.7833^{o}) \sin(76^{o}) \cos(2.0500^{o})$

$\cos \text{n} \approx 0.996703213875 \Rightarrow \text{n} \approx \cos^{-1}(0.996703213875) \Rightarrow \text{n} \approx 4.65374308576^{o}$.

Therefore, the approximate distance between the two capital cities is

$(4.65374308576^{o})\left({}^{\pi}/_{180^{o}}\right)(3959 \text{ miles})$

$\approx (0.081223139388)(3959 \text{ miles})$

≈ 322 miles.

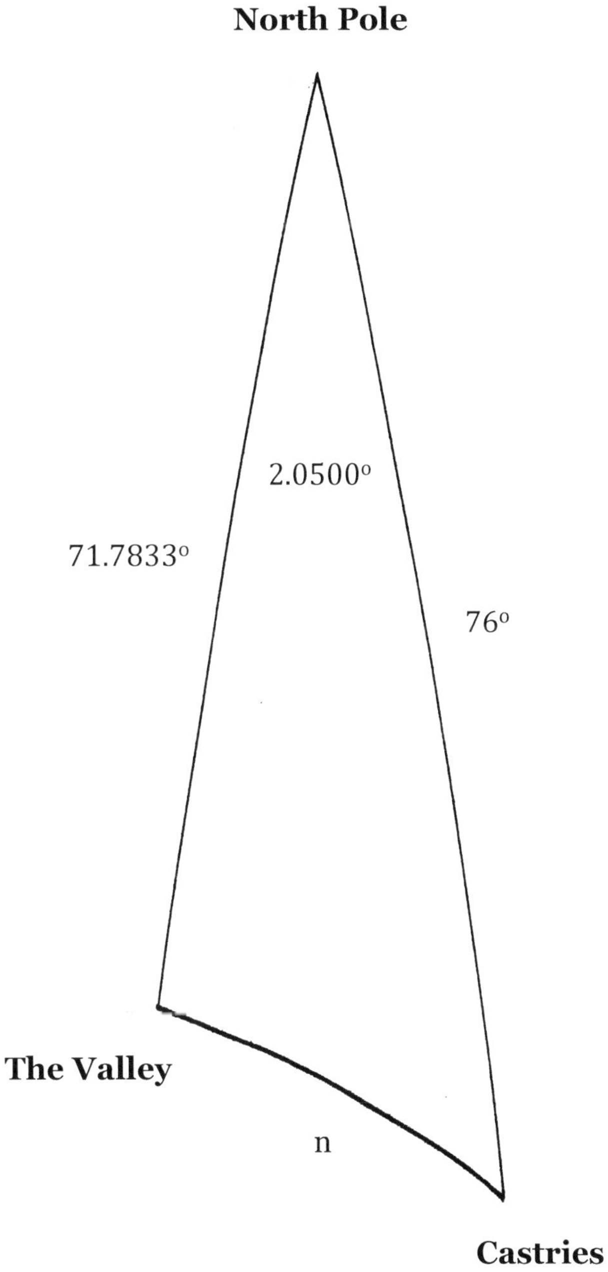
North Pole
2.0500°
71.7833°
76°
The Valley
n
Castries

Chapter 4
Problem 2
Solution

Oranjestad, Aruba is located at $12^o31' \text{ N} = \left(12 + {}^{31}/_{60}\right)^o \text{ N} \approx 12.5167^o \text{ N}$,

$70^o2' \text{ W} = \left(70 + {}^{2}/_{60}\right)^o \text{ W} \approx 70.0333^o \text{ W}$.

Port of Spain, Trinidad and Tobago is located at

$10^o39' \text{ N} = \left(10 + {}^{39}/_{60}\right)^o \text{ N} = 10.6500^o \text{ N}$, $61^o31' \text{ W} = \left(61 + {}^{31}/_{60}\right)^o \text{ W} \approx 61.5167^o \text{ W}$.

Two sides of the spherical triangle are $90^o - 12.5167^o = 77.4833^o$ and $90^o - 10.6500^o = 79.3500^o$, respectively. The angle at the North Pole is $70.0333^o - 61.5167^o = 8.5166^o$.

See the illustration on the next page.

It follows that,

$\cos n = \cos(77.4833^o)\cos(79.3500^o) + \sin(77.4833^o)\sin(79.3500^o)\cos(8.5166^o)$ cos $n \approx 0.988889826394 \Rightarrow n \approx \cos^{-1}(0.988889826394) \Rightarrow n \approx 8.54871748611^o$.

Therefore, the approximate distance between the two capital cities

$(8.54871748611^o)\left({}^{\pi}/_{180^o}\right)(3959 \text{ miles})$

$\approx (0.149203266955)(3959 \text{ miles})$

≈ 591 miles.

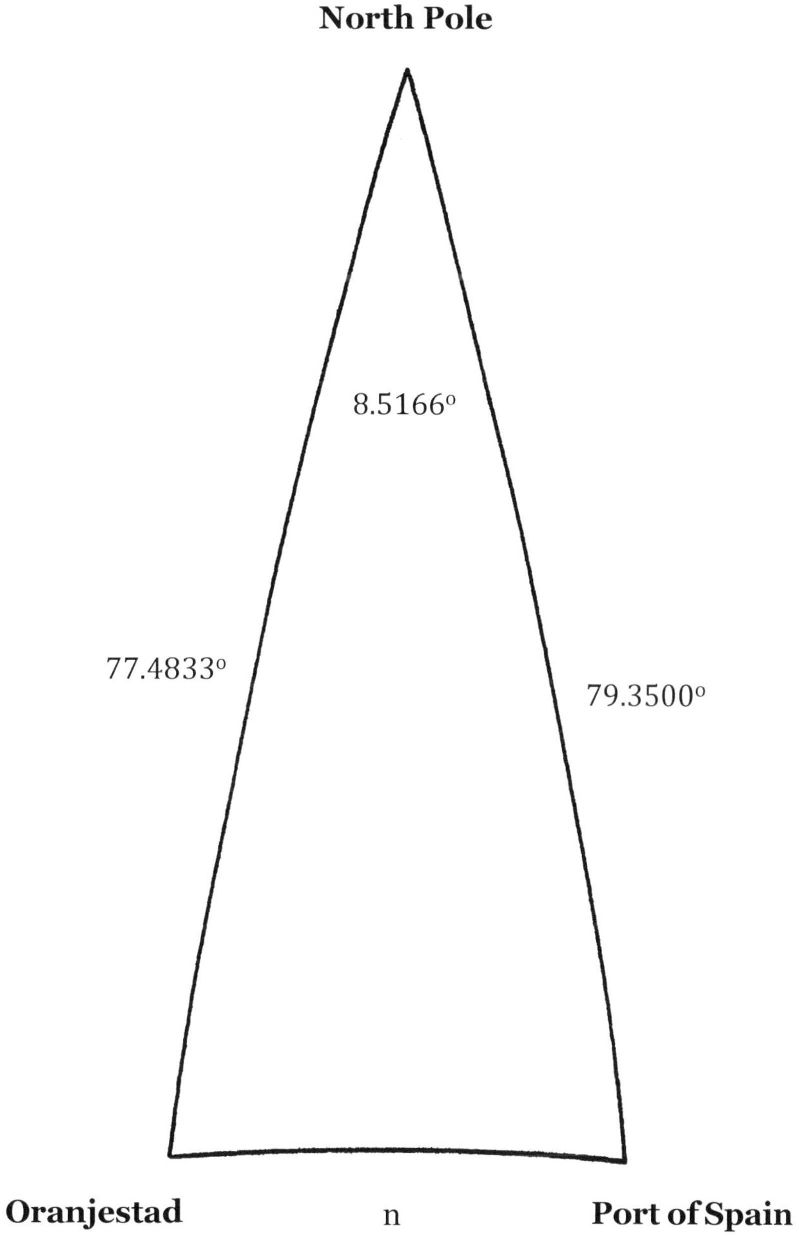
North Pole
8.5166°
77.4833°
79.3500°
Oranjestad
n
Port of Spain

Chapter 4

Problem 3

Solution

Belmopan, Belize is located at $17^o15'\text{N} = \left(17 + {}^{15}/_{60}\right)^o \text{N} = 17.2500^o\text{N}$,

$88^o46'\,\text{W} = \left(88 + {}^{46}/_{60}\right)^o \text{W} \approx 88.7667^o\,\text{W}$.

Kingston, Jamaica is located at $18^o0'\text{N} = 18^o\text{N}$,

$76^o48'\text{W} = \left(76 + {}^{48}/_{60}\right)^o \text{W} \approx 76.8000^o\text{W}$.

Two sides of the spherical triangle are $90^o - 17.2500^o = 72.7500^o$ and $90^o - 18^o = 72^o$, respectively. The angle at the North Pole is $88.7667^o - 76.8000^o = 11.9667^o$.

See the illustration on the next page.

It follows that

$\cos n = \cos(72.7500^o)\cos(72^o) + \sin(72.7500^o)\ \sin(72^o)\cos(11.9667^o)$

$\cos n \approx 0.980175879026 \Rightarrow\ n \approx \cos^{-1}(0.980175879026)\ \Rightarrow\ n \approx 11.427590797^o$.

Therefore, the approximate distance between the two capital cities is

$(11.427590797^o)\left({}^{\pi}/_{180^o}\right)(3959 \text{ miles})$

$\approx (0.199449084979)(3959 \text{ miles})$

$\approx\ 790\ \text{ miles}$.

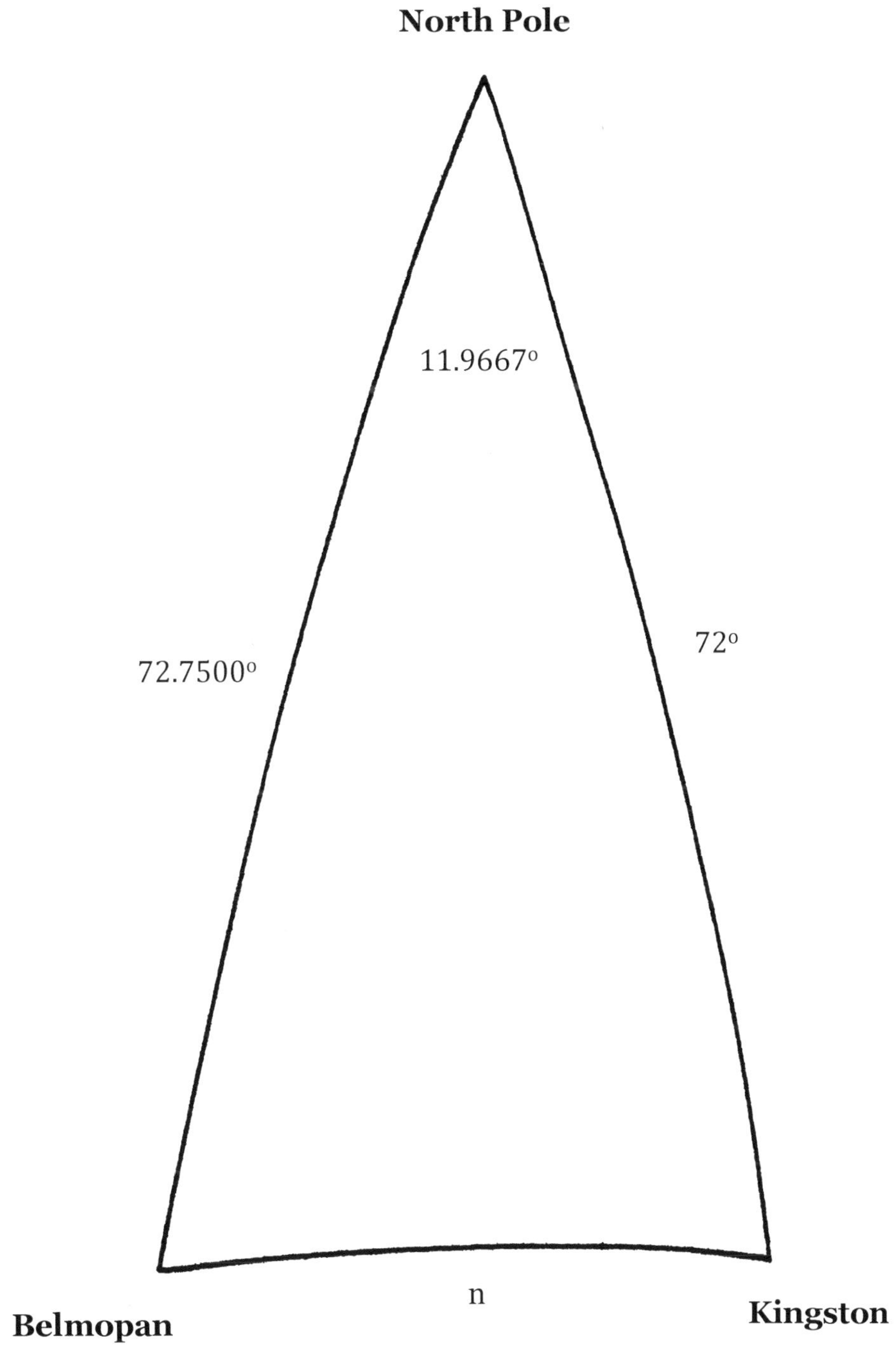
North Pole
11.9667°
72.7500°
72°
n
Belmopan
Kingston

Chapter 4
Problem 4
Solution

Havana, Cuba is located at $23^o7'\text{N} = \left(23 + {}^{7}/_{60}\right)^o \text{N} \approx 23.1167^o\text{N}$,

$82^o21'\text{W} = \left(82 + {}^{21}/_{60}\right)^o \text{W} \approx 82.3500^o\text{W}$.

Bridgetown, Barbados is located at $13^o6'\text{N} = \left(13 + {}^{6}/_{60}\right)^o \text{N} = 13.1000^o\text{N}$,

$59^o37'\text{W} = \left(59 + {}^{37}/_{60}\right)^o \text{W} = 59.6167^o\text{W}$.

Two sides of the spherical triangle are $90^o - 23.1167^o = 66.8833^o$ and $90^o - 13.1000^o = 76.9000^o$, respectively. The angle at the North Pole is $82.3500^o - 59.6167^o = 22.7333^o$.

See the illustration on the next page.

It follows that

$\cos n = \cos(66.8833^o)\cos(76.9000^o) + \sin(66.8833^o)\sin(76.9000^o)\cos(22.7333^o)$

$\cos n \approx 0.9151677904 \Rightarrow n \approx \cos^{-1}(0.9151677904) \Rightarrow n \approx 23.77043324^o$.

Therefore, the approximate distance between the two capital cities

$(23.77043324^o)\left({}^{\pi}/_{180^o}\right)(3959 \text{ miles})$

$\approx (0.4148723247)(3959 \text{ miles})$

$\approx$ 1,642 miles.

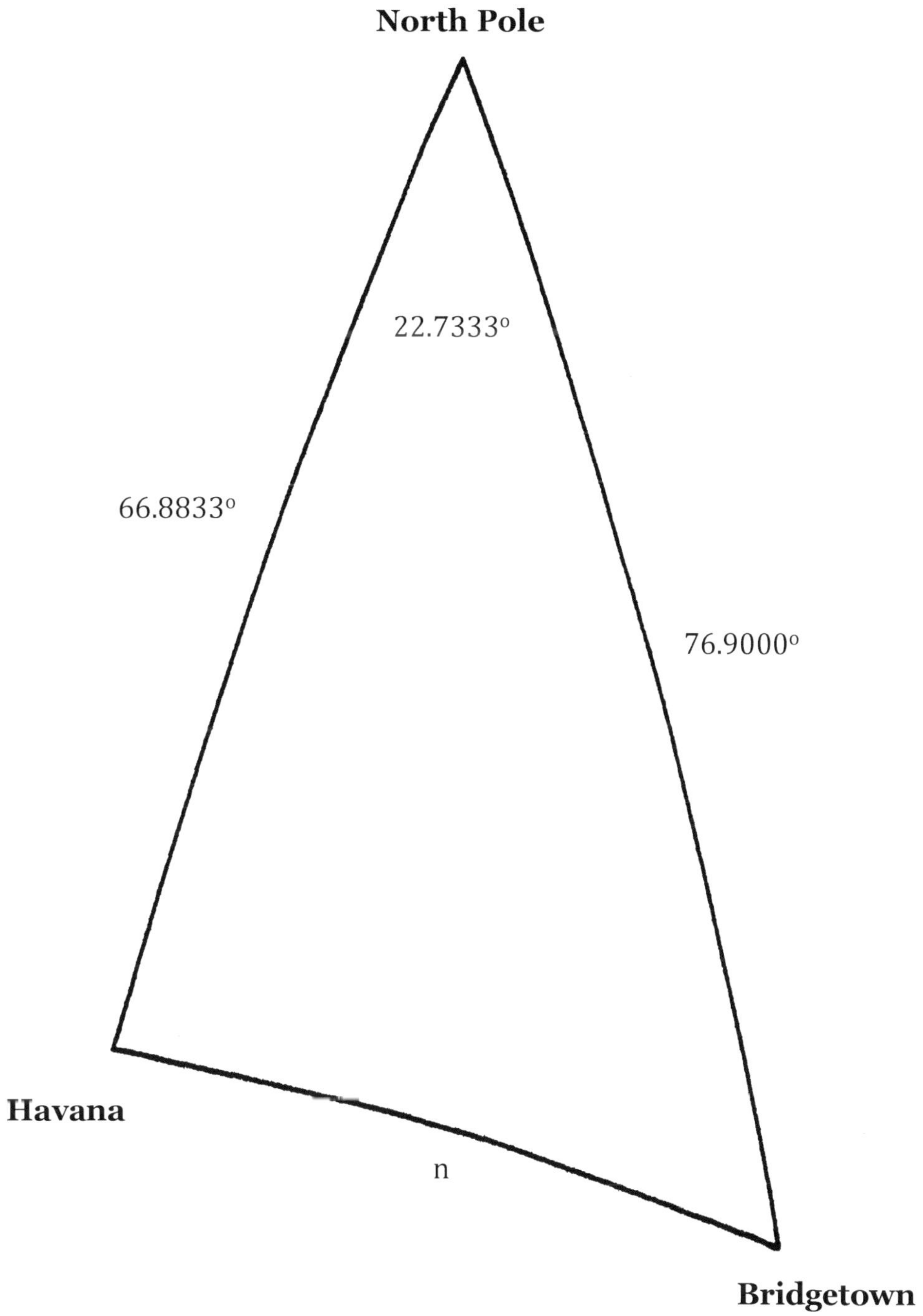
North Pole
22.7333°
66.8833°
76.9000°
Havana
n
Bridgetown

Chapter 4

Problem 5

Solution

Willemstad, Curacao is located at $12^o 6' \text{ N} = \left(12 + {}^{6}/_{60}\right)^o \text{ N} = 12.1000^o \text{ N}$,

$68^o 55' \text{ W} = \left(68 + {}^{55}/_{60}\right)^o \text{ W} \approx 68.9167^o \text{ W}$.

Saint Georges, Grenada is located at $12^o 3' \text{ N} = \left(12 + {}^{3}/_{60}\right)^o \text{ N} = 12.0500^o \text{ N}$,

$61^o 45' \text{ W} = \left(61 + {}^{45}/_{60}\right)^o \text{ W} = 61.7500^o \text{ W}$.

Two sides of the spherical triangle are $90^o - 12.1000^o = 77.9000^o$ and $90^o - 12.0500^o = 77.9500^o$, respectively. The angle at the North Pole is $68.9167^o - 61.7500^o = 7.1667^o$.

See the illustration on the next page.

It follows that

$\cos n = \cos(77.9000^o)\cos(77.9500^o) + \sin(77.9000^o)\ \sin(77.9500^o)\cos(7.1667^o)$

$\cos n \approx 0.99252888765 \Rightarrow n \approx \cos^{-1}(0.99252888765) \Rightarrow n \approx 7.00811183416^o$.

Therefore, the approximate distance between the two capital cities is

$(7.00811183416^o)\left({}^{\pi}/_{180^o}\right)(3959 \text{ miles})$

$\approx (0.122314625854)(3959 \text{ miles})$

$\approx$ 484 miles.

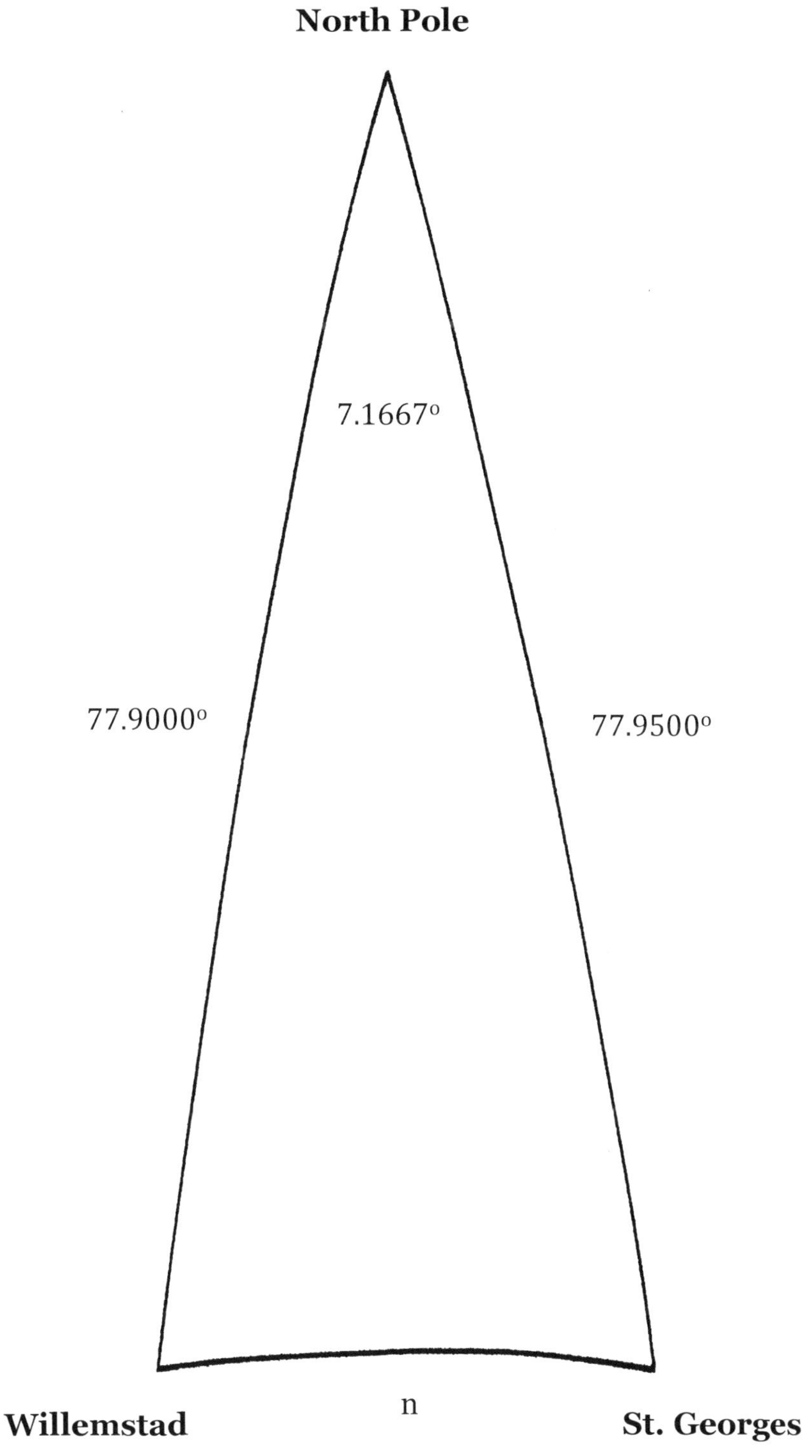
North Pole
7.1667°
77.9000°
77.9500°
n
Willemstad
St. Georges

Chapter 4
Problem 6
Solution

San Salvador, El Salvador is located at $13^o 42' \text{ N} = \left(13 + {}^{42}/_{60}\right)^o \text{ N} = 13.7000^o \text{ N}$,

$89^o 12' \text{ W} = \left(89 + {}^{12}/_{60}\right)^o \text{ W} = 89.2000^o \text{ W}$.

Road Town, British Virgin Islands is located at

$18^o 25' \text{ N} = \left(18 + {}^{25}/_{60}\right)^o \text{ N} \approx 18.4167^o \text{ N}$,

$64^o 37' \text{ W} = \left(64 + {}^{37}/_{60}\right)^o \text{ W} \approx 64.6167^o \text{ W}$.

Two sides of the spherical triangle are $90^o - 13.7000^o = 76.3000^o$ and $90^o - 18.4167^o = 71.5833^o$, respectively. The angle at the North Pole is $89.2000^o - 64.6167^o = 24.5833^o$.

See the illustration on the next page.

It follows that

$\cos n = \cos(76.3000^o)\cos(71.5833^o) + \sin(76.3000^o)\ \sin(71.5833^o)\cos(24.5833^o)$

$\cos n \approx 0.913060002226 \ \Rightarrow \ n \approx \cos^{-1}(0.913060002226) \ \Rightarrow \ n \approx 24.0682932593^o$.

Therefore, the approximate distance between the two capital cities is

$(24.0682932593^o)\left({}^{\pi}/_{180^o}\right)(3959 \text{ miles})$

$\approx (0.420070962711)(3959 \text{ miles})$

$\approx 1{,}663$ miles.

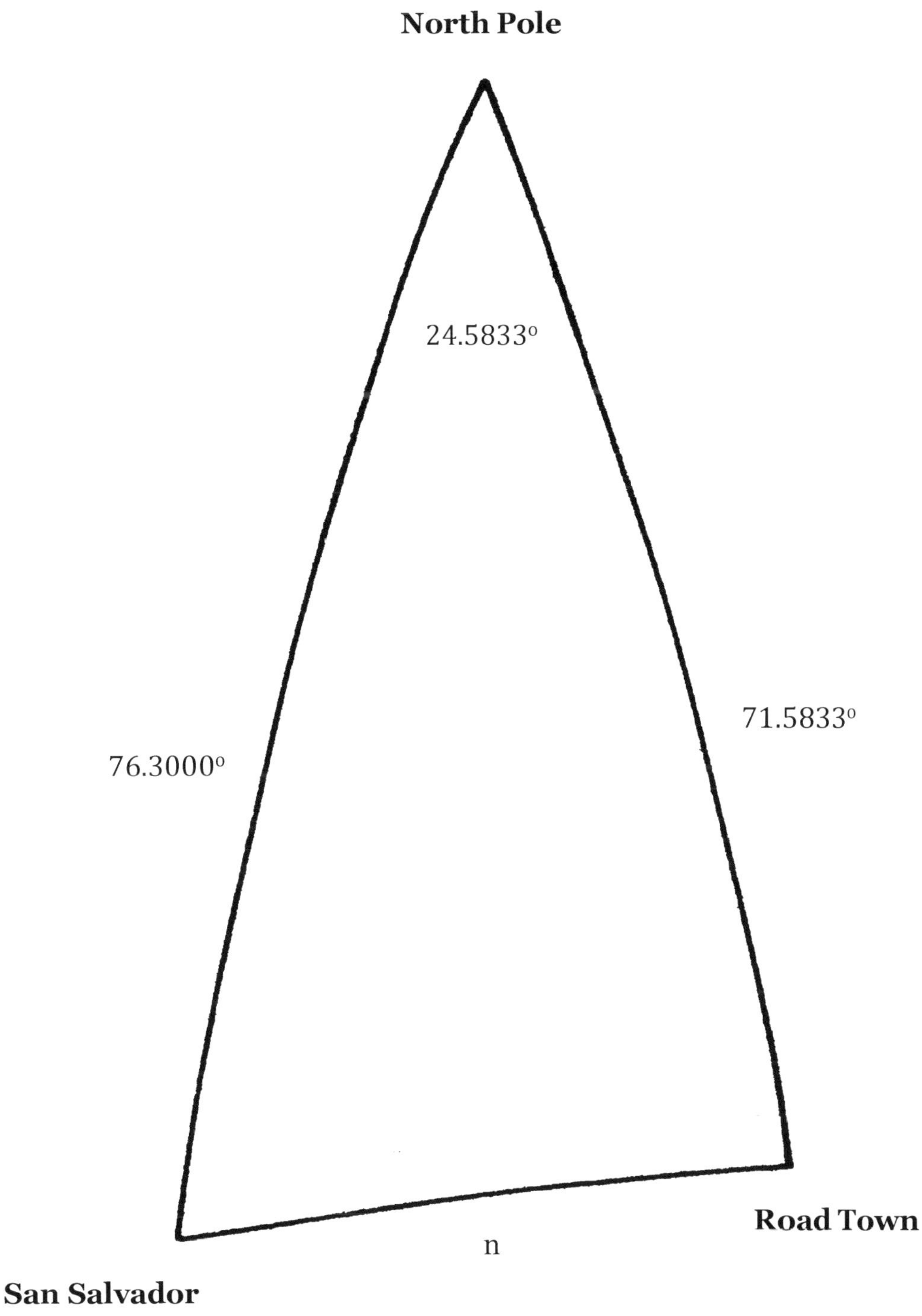
North Pole
24.5833°
76.3000°
71.5833°
Road Town
n
San Salvador

Chapter 4
Problem 7
Solution

Tegucigalpa, Honduras is located at $14^o6'\ \mathrm{N} = \left(14 + {}^{6}/_{60}\right)^o \mathrm{N} = 14.1000^o\mathrm{N}$,

$87^o13'\ \mathrm{W} = \left(87 + {}^{13}/_{60}\right)^o \mathrm{W} \approx 87.2167^o\mathrm{W}$.

San Jose, Costa Rica is located at $9^o56'\ \mathrm{N} = \left(9 + {}^{56}/_{60}\right)^o \mathrm{N} \approx 9.9333^o\mathrm{N}$,

$84^o5'\ \mathrm{W} = \left(84 + {}^{5}/_{60}\right)^o \mathrm{W} \approx 84.0833^o\mathrm{W}$.

Two sides of the spherical triangle are $90^o - 14.1000^o = 75.9000^o$ and $90^o - 9.9333^o = 80.0667^o$, respectively. The angle at the North Pole is $87.2167^o - 84.0833^o = 3.1334^o$.

See the illustration on the next page.

It follows that

$\cos \mathrm{n} = \cos(75.9000^o)\cos(80.0667^o) + \sin(75.9000^o)\ \sin(80.0667^o)\cos(3.1334^o)$

$\cos \mathrm{n} \approx 0.995928628795 \ \Rightarrow \mathrm{n} \approx \cos^{-1}(0.995928628795) \ \Rightarrow \mathrm{n} \approx 5.17196334862^o$.

Therefore, the approximate distance between the two capital cities

$(5.17196334862^o)\left({}^{\pi}/_{180^o}\right)(3959 \text{ miles})$

$\approx (0.090267789226)(3959 \text{ miles})$

$\approx\ 357 \text{ miles}$.

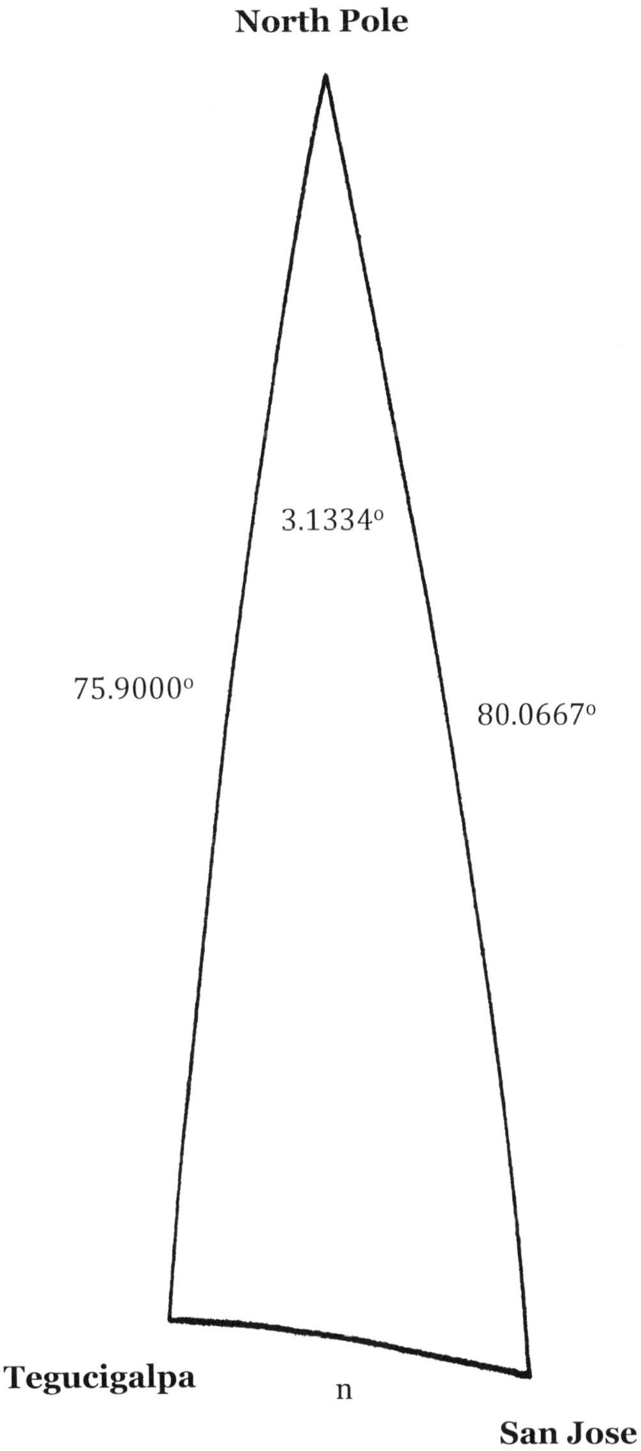
North Pole
3.1334°
75.9000°
80.0667°
Tegucigalpa
n
San Jose

Chapter 4
Problem 8
Solution

Plymouth, Montserrat is located at $16^o 42' \text{ N} = \left(16 + {}^{42}/_{60}\right)^o \text{ N} = 16.7000^o \text{ N}$, $62^o 13' \text{ W} = \left(62 + {}^{13}/_{60}\right)^o \text{ W} \approx 62.2167^o \text{ W}$.

Roseau, Dominica is located at $15^o 18' \text{ N} = \left(15 + {}^{18}/_{60}\right)^o \text{ N} = 15.3000^o \text{ N}$, $61^o 24' \text{ W} = \left(61 + {}^{24}/_{60}\right)^o \text{ W} = 61.4000^o \text{ W}$.

Two sides of the spherical triangle are $90^o - 16.7000^o = 73.3000^o$ and $90^o - 15.3000^o = 74.7000^o$, respectively. The angle at the North Pole is $62.2167^o - 61.4000^o = 0.8167^o$.

See the illustration on the next page.

It follows that

$\cos n = \cos(73.3000^o)\cos(74.7000^o) + \sin(73.3000^o)\sin(74.7000^o)\cos(0.8167^o)$

$\cos n \approx 0.99960763517 \Rightarrow n \approx \cos^{-1}(0.99960763517) \Rightarrow n \approx 1.60508068087^o$.

Therefore, the approximate distance between the two capital cities is

$(1.60508068087^o)\left({}^{\pi}/_{180^o}\right)(3959 \text{ miles})$

$\approx (0.028013942641)(3959 \text{ miles})$

≈ 111 miles.

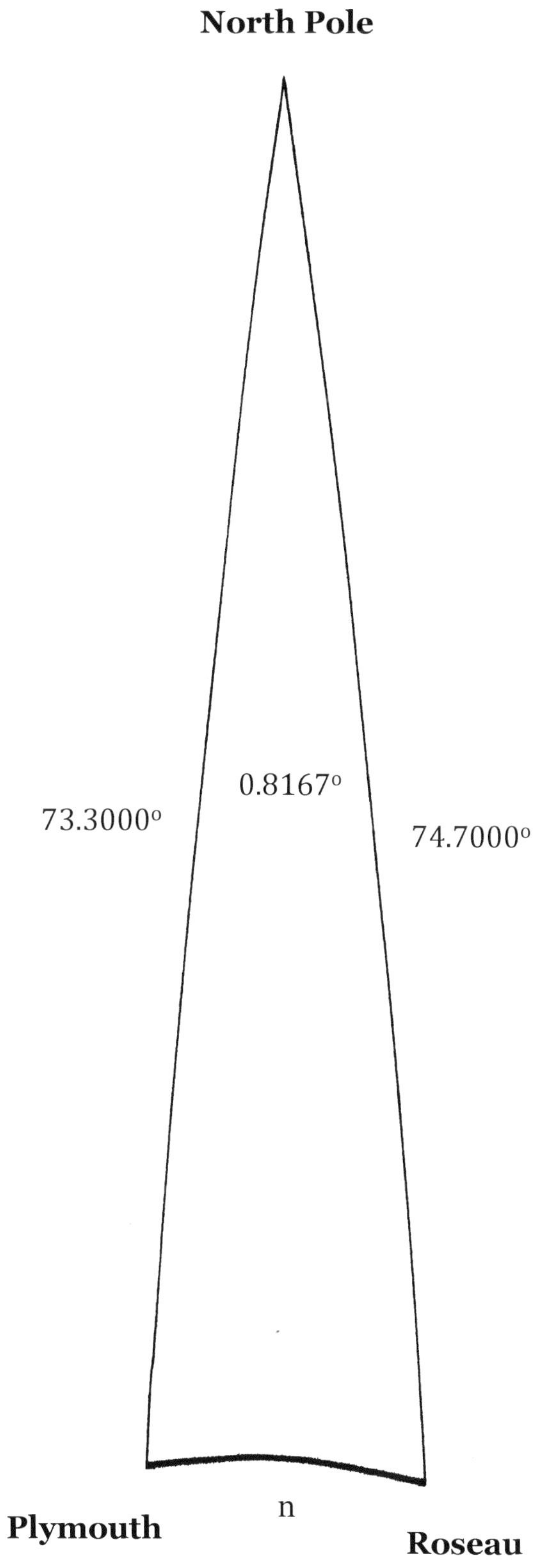
North Pole
73.3000°
0.8167°
74.7000°
Plymouth
n
Roseau

Chapter 4
Problem 9
Solution

Panama City, Panama is located at $8^o58'$ N $= \left(8 + {}^{58}/_{60}\right)^o$ N $\approx 8.9667^o$ N , $79^o32'$ W $= \left(79 + {}^{32}/_{60}\right)^o$ W $\approx 79.5333^o$ W .

Nassau, The Bahamas is located at $25^o5'$ N $= \left(25 + {}^{5}/_{60}\right)^o$ N $\approx 25.0833^o$ N , $77^o21'$ W $= \left(77 + {}^{21}/_{60}\right)^o$ W $= 77.3500^o$ W .

Two sides of the spherical triangle are $90^o - 8.9667^o = 81.0333^o$ and $90^o - 25.0833^o = 64.9167^o$, respectively. The angle at the North Pole is $79.5333^o - 77.3500^o = 2.1833^o$.

See the illustration on the next page.

It follows that

$\cos n = \cos(81.0333^o)\cos(64.9167^o) + \sin(81.0333^o)\ \sin(64.9167^o)\cos(2.1833^o)$

$\cos n \approx 0.960049328139 \ \Rightarrow\ n \approx \cos^{-1}(0.960049328139) \ \Rightarrow\ n \approx 16.2501077502^o$.

Therefore, the approximate distance between the two capital cities is

$(16.2501077502^o)\left({}^{\pi}/_{180^o}\right)(3959 \text{ miles})$

$\approx (0.283617884044)(3959 \text{ miles})$

$\approx\ 1{,}123\ \text{ miles}.$

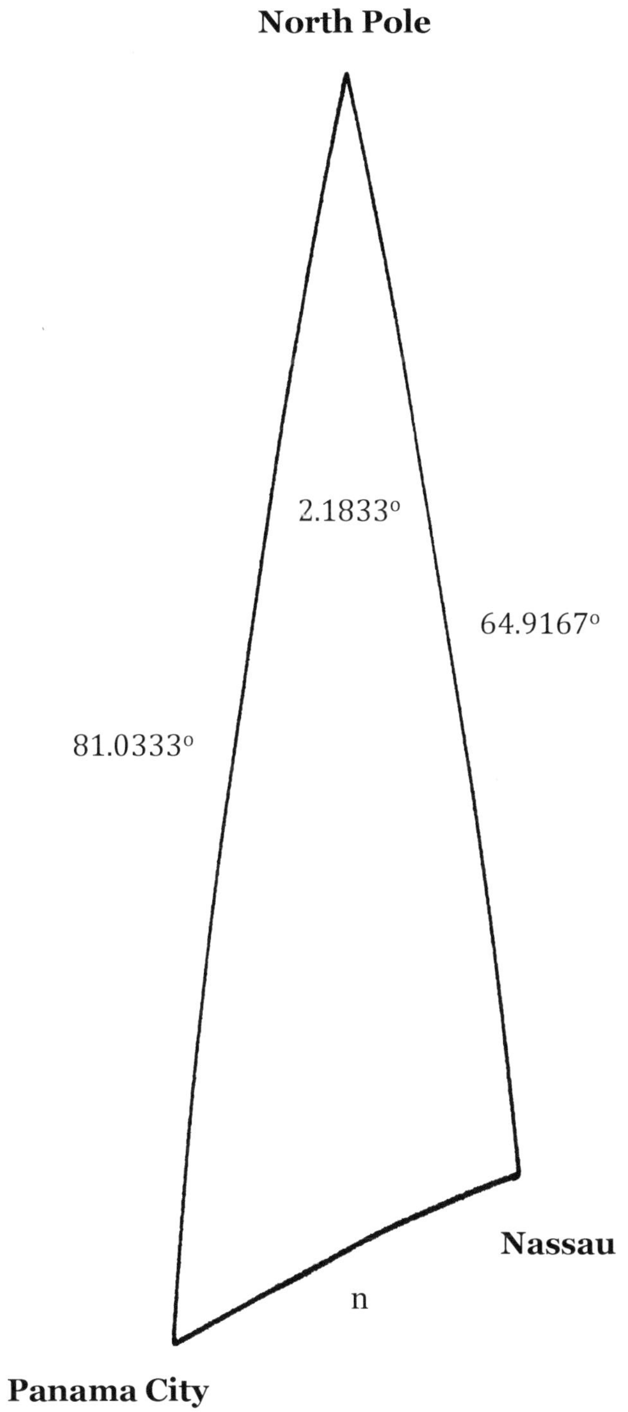
North Pole
2.1833°
64.9167°
81.0333°
Nassau
n
Panama City

Chapter 4

Problem 10

Solution

Nassau, The Bahamas is located at $25^\circ 5'\text{ N} = \left(25 + {}^{5}/_{60}\right)^\circ \text{ N} \approx 25.0833^\circ \text{ N}$,

$77^\circ 21'\text{ W} = \left(77 + {}^{21}/_{60}\right)^\circ \text{ W} = 77.3500^\circ \text{ W}$.

San Juan, Puerto Rico is located at $18^\circ 28'\text{ N} = \left(18 + {}^{28}/_{60}\right)^\circ \text{ N} \approx 18.4667^\circ \text{ N}$,

$66^\circ 7'\text{ W} = \left(66 + {}^{7}/_{60}\right)^\circ \text{ W} \approx 66.1167^\circ \text{ W}$.

Two sides of the spherical triangle are $90^\circ - 25.0833^\circ = 64.9167^\circ$ and $90^\circ - 18.4667^\circ = 71.5333^\circ$, respectively. The angle at the North Pole is $77.3500^\circ - 66.1167^\circ = 11.2333^\circ$.

See the illustration on the next page.

It follows that

$\cos n = \cos(71.5333^\circ)\cos(64.9167^\circ) + \sin(71.5333^\circ)\ \sin(64.9167^\circ)\cos(11.2333^\circ)$

$\cos n \approx 0.976881708497 \Rightarrow\ n \approx \cos^{-1}(0.976881708497)\ \Rightarrow\ n \approx 12.3439952458^\circ$.

Therefore, the approximate distance between the two capital cities is

$(12.3439952458^\circ)\left({}^{\pi}/_{180^\circ}\right)(3959\text{ miles})$

$\approx (0.215443359889)(3959\text{ miles})$

$\approx\ 853\text{ miles.}$

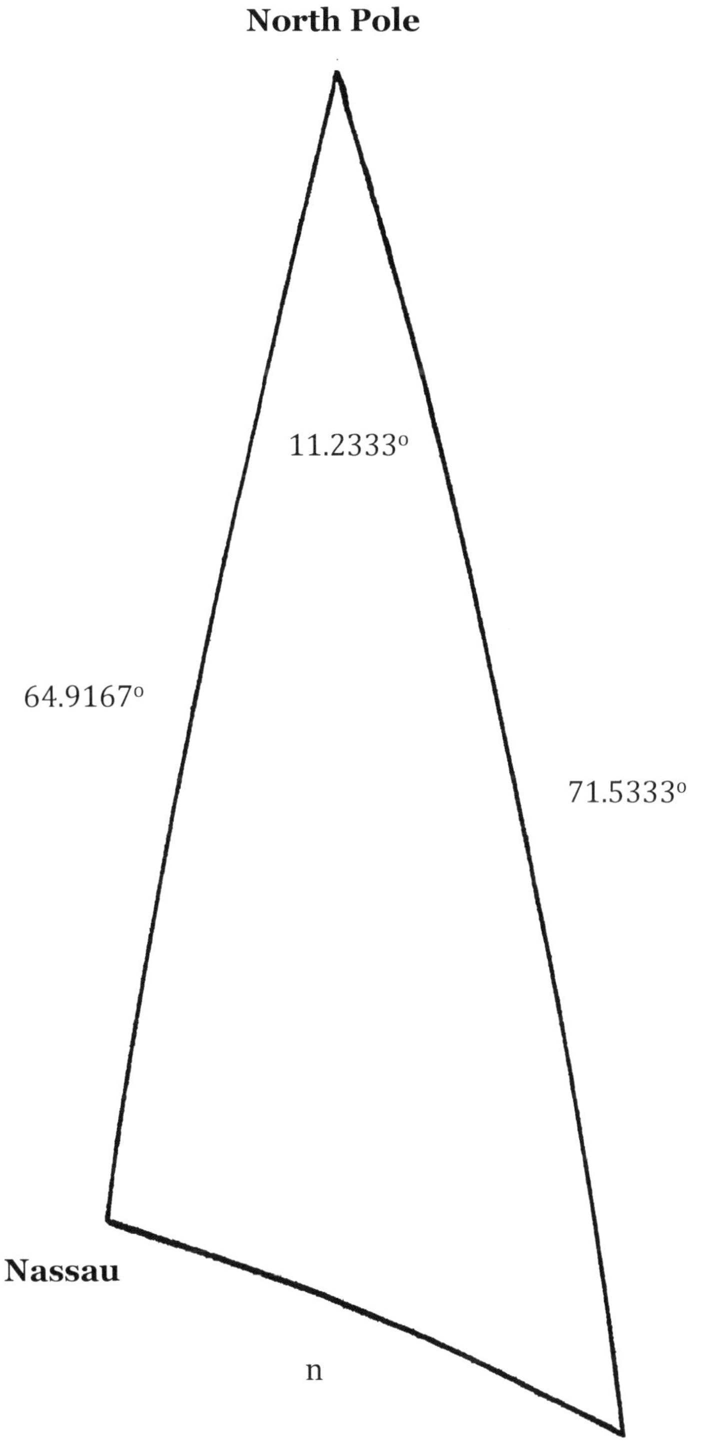
North Pole
11.2333°
64.9167°
71.5333°
Nassau
n
San Juan

Chapter 4
Problem 11
Solution

Marigot, Saint Martin is located at $18^{o}4' \text{ N} = \left(18 + {}^{4}/_{60}\right)^{o} \text{ N} \approx 18.0667^{o} \text{ N}$,

$63^{o}5' \text{ W} = \left(63 + {}^{5}/_{60}\right)^{o} \text{ W} \approx 63.0833^{o} \text{ W}$.

Gustavia, Saint Barthelemy is located at

$17^{o}53' \text{ N} = \left(17 + {}^{53}/_{60}\right)^{o} \text{ N} \approx 17.8833^{o} \text{ N}$,

$62^{o}51' \text{ W} = \left(62 + {}^{51}/_{60}\right)^{o} \text{ W} = 62.8500^{o} \text{ W}$.

Two sides of the spherical triangle are $90^{o} - 18.0667^{o} \approx 71.9333^{o}$ and $90^{o} - 17.8833^{o} \approx 72.1167^{o}$, respectively. The angle at the North Pole is $63.0833^{o} - 62.8500^{o} \approx 0.2333^{o}$.

See the illustration on the next page.

It follows that

$\cos n = \cos(71.9333^{o}) \cos(72.1167^{o}) + \sin(71.9333^{o}) \sin(72.1167^{o}) \cos(0.2333^{o})$

$\cos n \approx 0.999987376552 \Rightarrow n \approx \cos^{-1}(0.999987376552\ ) \Rightarrow n \approx 0.28789033094^{o}$.

Therefore, the approximate distance between the two capital cities is

$(0.28789033094^{o})\left({}^{\pi}/_{180^{o}}\right)(3959 \text{ miles})$

$\approx (0.00502463416)(3959 \text{ miles})$

$\approx 20 \text{ miles}$.

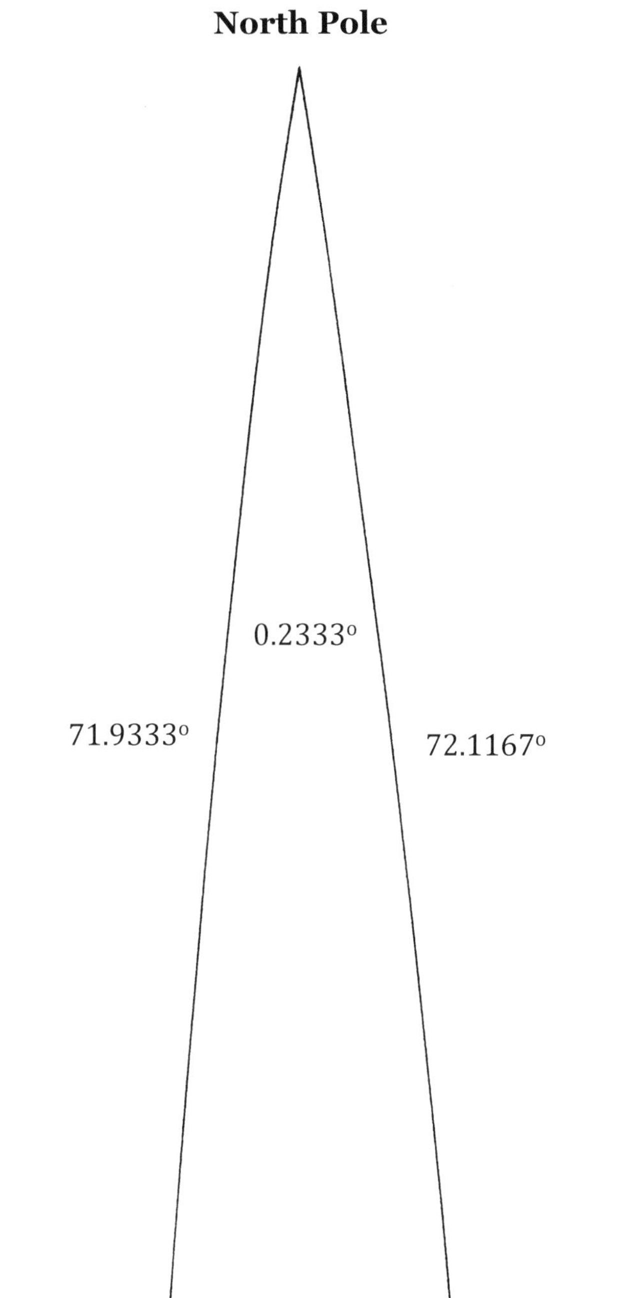
North Pole
0.2333°
71.9333°
72.1167°
Marigot
n
Gustavia

Chapter 4
Problem 12
Solution

Philipsburg, Sint Maarten is located at $18^{o}1'\ \mathrm{N} = \left(18 + {}^{1}/_{60}\right)^{o}\ \mathrm{N} \approx 18.0167^{o}\ \mathrm{N}$,

$63^{o}2'\ \mathrm{W} = \left(63 + {}^{2}/_{60}\right)^{o}\ \mathrm{W} \approx 63.0333^{o}\ \mathrm{W}$.

Saint John's, Antigua and Barbuda is located at

$17^{o}7'\ \mathrm{N} = \left(17 + {}^{7}/_{60}\right)^{o}\ \mathrm{N} \approx 17.1167^{o}\ \mathrm{N}$, $61^{o}51'\ \mathrm{W} = \left(61 + {}^{51}/_{60}\right)^{o}\ \mathrm{W} = 61.8500^{o}\ \mathrm{W}$.

Two sides of the spherical triangle are $90^{o} - 18.0167^{o} = 71.9833^{o}$ and $90^{o} - 17.1167^{o} = 72.8833^{o}$, respectively. The angle at the North Pole is $63.0333^{o} - 61.8500^{o} = 1.1833^{o}$.

See the illustration on the next page.

It follows that

$\cos \mathrm{n} = \cos(71.9833^{o}) \cos(72.8833^{o}) + \sin(71.9833^{o}) \sin(72.8833^{o}) \cos(1.1833^{o})$

$\cos \mathrm{n} \approx 0.999682816718 \Rightarrow \mathrm{n} \approx \cos^{-1}(0.999682816718) \Rightarrow \mathrm{n} \approx 1.44312601147^{o}$.

Therefore, the approximate distance between the two capital cities

$(1.44312601147^{o}) \left({}^{\pi}/_{180^{o}}\right) (3959 \text{ miles})$

$\approx (0.025187300421)(3959 \text{ miles})$

$\approx 100 \text{ miles}$.

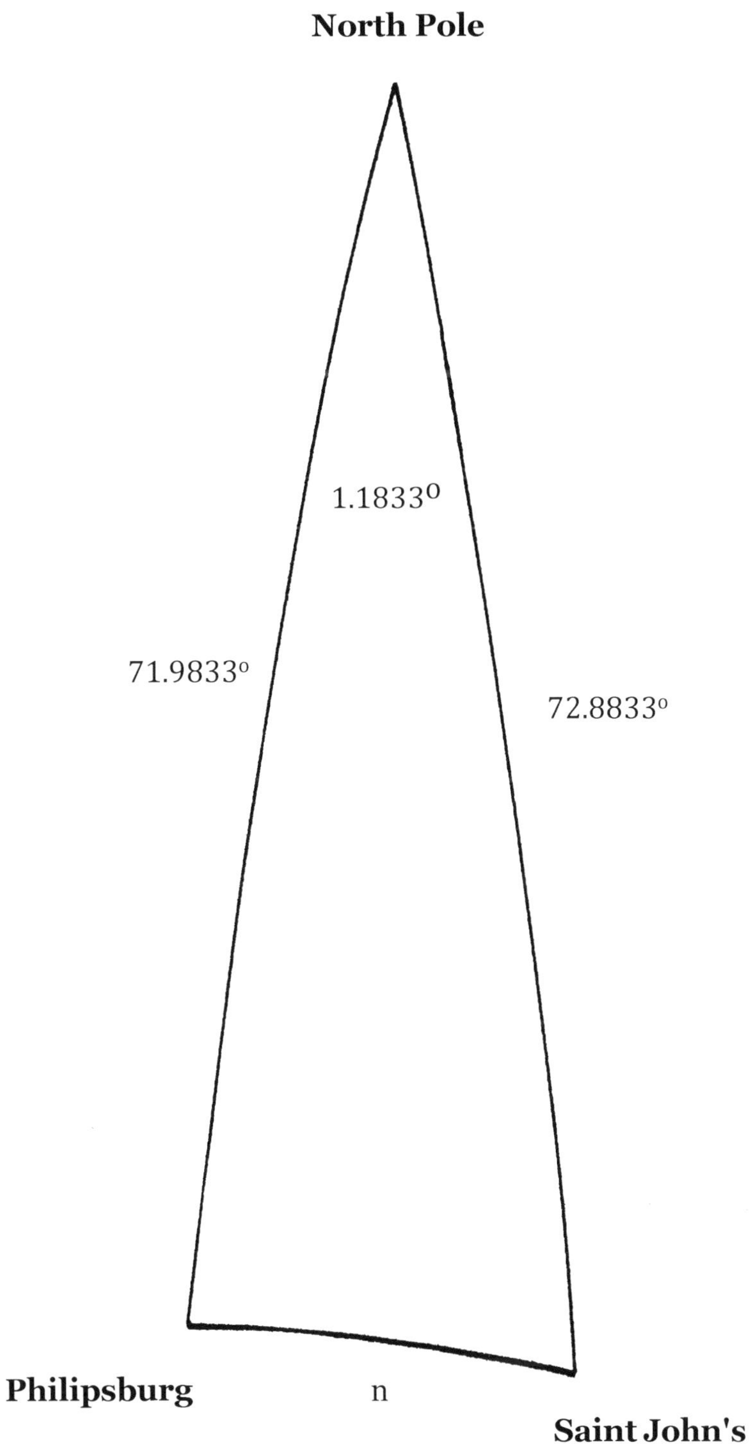
North Pole
1.1833°
71.9833°
72.8833°
Philipsburg
n
Saint John's

Chapter 4
Problem 13
Solution

Grand Turk, Turks and Caicos Islands is located at

$21^o 28' \text{ N} = \left(21 + {}^{28}/_{60}\right)^o \text{ N} \approx 21.4667^o \text{ N}$,

$71^o 8' \text{ W} = \left(71 + {}^{8}/_{60}\right)^o \text{ W} \approx 71.1333^o \text{ W}$.

Santo Domingo, Dominican Republic is located at

$18^o 28' \text{ N} = \left(18 + {}^{28}/_{60}\right)^o \text{ N} \approx 18.4667^o \text{ N}$,

$69^o 54' \text{ W} = \left(69 + {}^{54}/_{60}\right)^o \text{ W} = 69.9000^o \text{ W}$.

Two sides of the spherical triangle are $90^o - 21.4667^o = 68.5333^o$ and $90^o - 18.4667^o = 71.5333^o$, respectively. The angle at the North Pole is $71.1333^o - 69.9000^o = 1.2333^o$.

See the illustration on the next page.

It follows that

$\cos n = \cos(68.5333^o)\cos(71.5333^o) + \sin(68.5333^o)\ \sin(71.5333^o)\cos(1.2333^o)$

$\cos n \approx 0.998425048728 \Rightarrow n \approx \cos^{-1}(0.998425048728\ ) \Rightarrow n \approx 3.21609009735^o$.

Therefore, the approximate distance between the two capital cities is

$(3.21609009735^o)\left({}^{\pi}/_{180^o}\right)(3959 \text{ miles})$

$\approx (0.05613136124)(3959 \text{ miles})$

≈ 222 miles.

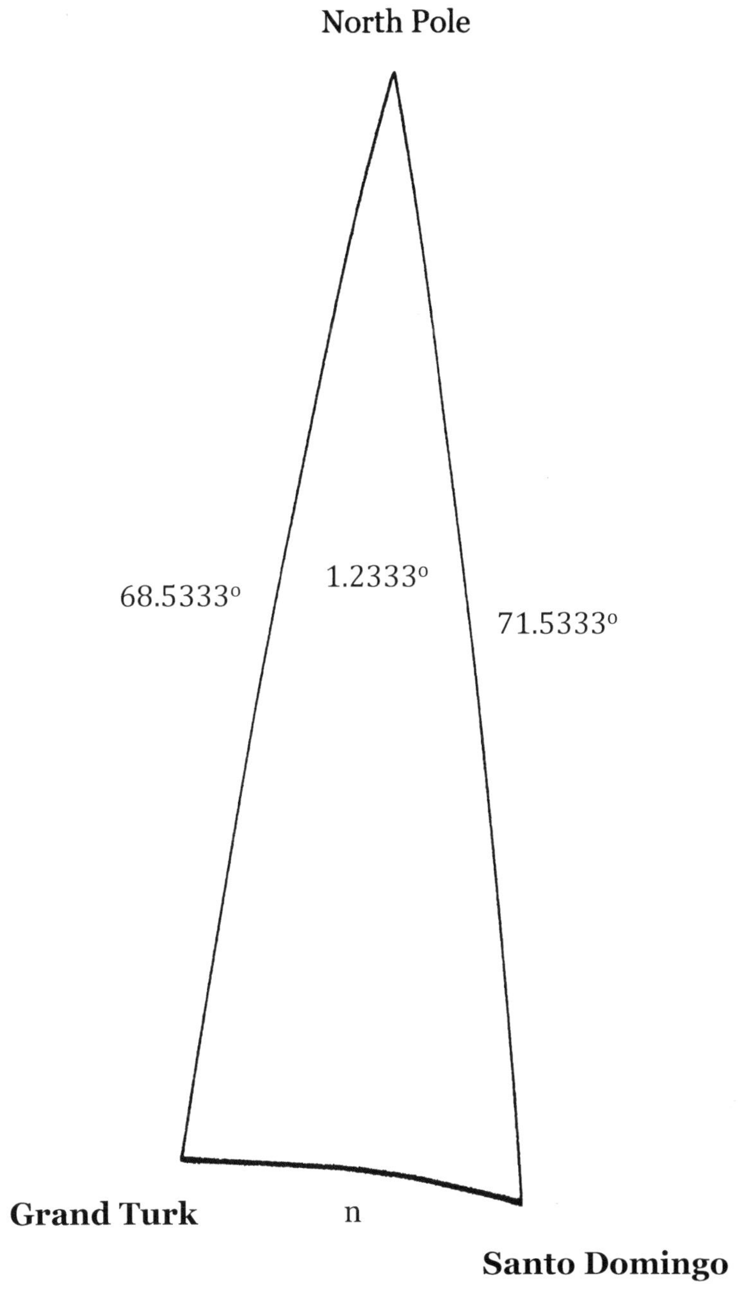
North Pole
68.5333°
1.2333°
71.5333°
Grand Turk
n
Santo Domingo

Chapter 4
Problem 14
Solution

Charlotte Amalie, U.S. Virgin Islands is located at

$18^{o}21'\text{ N} = \left(18 + {}^{21}/_{60}\right)^{o}\text{ N} = 18.3500^{o}\text{ N}$,

$64^{o}56'\text{ W} = \left(64 + {}^{56}/_{60}\right)^{o}\text{ W} \approx 64.9333^{o}\text{ W}$.

Basseterre, Saint Kitts and Nevis is located at

$17^{o}18'\text{ N} = \left(17 + {}^{18}/_{60}\right)^{o}\text{ N} = 17.3000^{o}\text{ N}$,

$62^{o}43'\text{ W} = \left(62 + {}^{43}/_{60}\right)^{o}\text{ W} \approx 62.7167^{o}\text{ W}$.

Two sides of the spherical triangle are $90^{o} - 18.3500^{o} = 71.6500^{o}$ and $90^{o} - 17.3000^{o} = 72.7000^{o}$, respectively. The angle at the North Pole is $64.9333^{o} - 62.7167^{o} = 2.2166^{o}$.

See the illustration on the next page.

It follows that

$\cos\text{n} = \cos(71.6500^{o})\cos(72.7000^{o}) + \sin(71.6500^{o})\ \sin(72.7000^{o})\cos(2.2166^{o})$

$\cos\text{n} \approx 0.999154013352 \Rightarrow \text{n} \approx \cos^{-1}(0.999154013352\ ) \Rightarrow \text{n} \approx 2.35694803074^{o}$.

Therefore, the approximate distance between the two capital cities is

$(2.35694803074^{o})\left({}^{\pi}/_{180^{o}}\right)(3959\text{ miles})$

$\approx (0.041136503435)(3959\text{ miles})$

$\approx 163\text{ miles}$.

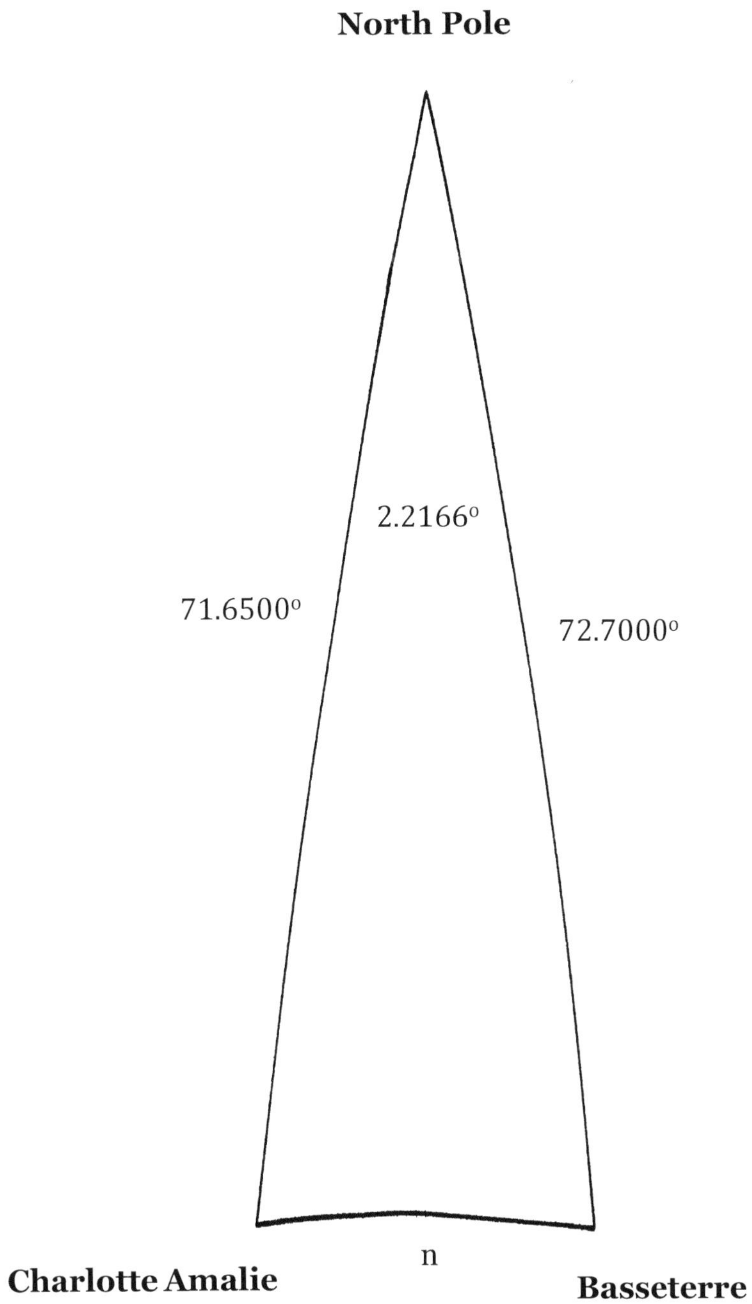
North Pole
2.2166°
71.6500°
72.7000°
n
Charlotte Amalie
Basseterre

Chapter 5

Central Asia

Example 1

Ashgabat, Turkmenistan is located at $37^o57' \text{ N} = \left(37 + {}^{57}/_{60}\right)^o \text{ N} = 37.9500^o \text{ N}$, $58^o23' \text{ E} = \left(58 + {}^{23}/_{60}\right)^o \text{ E} \approx 58.3833^o \text{ E}$.

Tashkent, Uzbekistan is located at $41^o19' \text{ N} = \left(41 + {}^{19}/_{60}\right)^o \text{ N} \approx 41.3167^o \text{ N}$, $69^o15' \text{ E} = \left(69 + {}^{15}/_{60}\right)^o \text{ E} = 69.2500^o \text{ E}$.

Two sides of the spherical triangle are $90^o - 37.9500^o = 52.0500^o$ and $90^o - 41.3167^o = 48.6833^o$, respectively. The angle at the North Pole is $69.2500^o - 58.3833^o = 10.8667^o$.

See the illustration on the next page.

It follows that

$\cos n = \cos(52.0500^o)\cos(48.6833^o) + \sin(52.0500^o)\ \sin(48.6833^o)\cos(10.8667^o)$

$\cos n \approx 0.987654060454 \;\Rightarrow\; n \approx \cos^{-1}(0.987654060454) \;\Rightarrow\; n \approx 9.01254679131^o$.

Therefore, the approximate distance between the two capital cities is
$(9.01254679131^o)\left({}^{\pi}/_{180^o}\right)(3959 \text{ miles})$

$\approx (0.157298615498)(3959 \text{ miles})$

≈ 623 miles.

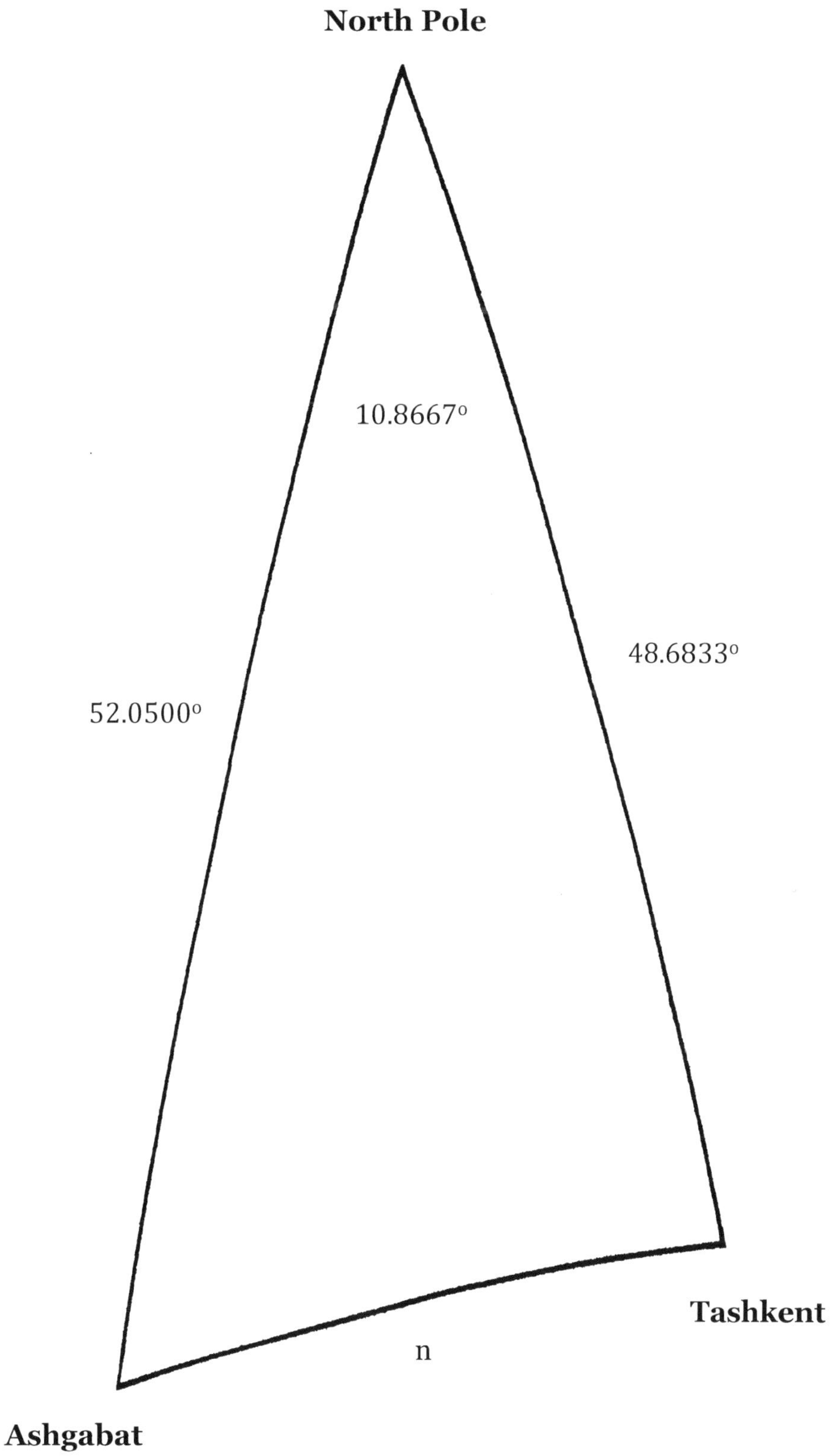
North Pole
10.8667°
52.0500°
48.6833°
Tashkent
n
Ashgabat

Chapter 5 Problems:

Approximate the distance between the following capitals in Central Asia.

1. Astana, Kazakhstan ($51^o10'$ N $71^o25'$ E)
   to Bishkek, Kyrgyzstan ($42^o52'$ N $74^o36'$ E)

2. Moscow, Russia ($55^o45'$ N $37^o36'$ E)
   to Dushanbe, Tajikistan ($38^o33'$ N $68^o46'$ E)

Chapter 5
Problem 1
Solution

Astana, Kazakhstan is located at $51^o10' \text{ N} = \left(51 + {}^{10}/_{60}\right)^o \text{ N} \approx 51.1667^o \text{ N}$,
$71^o25' \text{ E} = \left(71 + {}^{25}/_{60}\right)^o \text{ E} \approx 71.4167^o \text{ E}$.

Bishkek, Kyrgyzstan is located at $42^o52' \text{ N} = \left(42 + {}^{52}/_{60}\right)^o \text{ N} \approx 42.8667^o \text{ N}$,
$74^o36' \text{ E} = \left(74 + {}^{36}/_{60}\right)^o \text{ E} \approx 74.6000^o \text{ E}$.

Two sides of the spherical triangle are $90^o - 51.1667^o = 38.8333^o$ and $90^o - 42.8667^o = 47.1333^o$, respectively. The angle at the North Pole is $74.6000^o - 71.4167^o = 3.1833^o$.

See the illustration on the next page.

It follows that

$\cos n = \cos(38.8333^o)\cos(47.1333^o) + \sin(38.8333^o)\ \sin(47.1333^o)\cos(3.1833^o)$

$\cos n \approx 0.988816632028 \ \Rightarrow\ n \approx \cos^{-1}(0.988816632028) \ \Rightarrow\ n \approx 8.57688347242^o$.

Therefore, the approximate distance between the two capital cities is
$(8.57688347242^o)\left({}^{\pi}/_{180^o}\right)(3959)$

$\approx (0.149694856154)(3959 \text{ miles})$

$\approx\ 593$ miles.

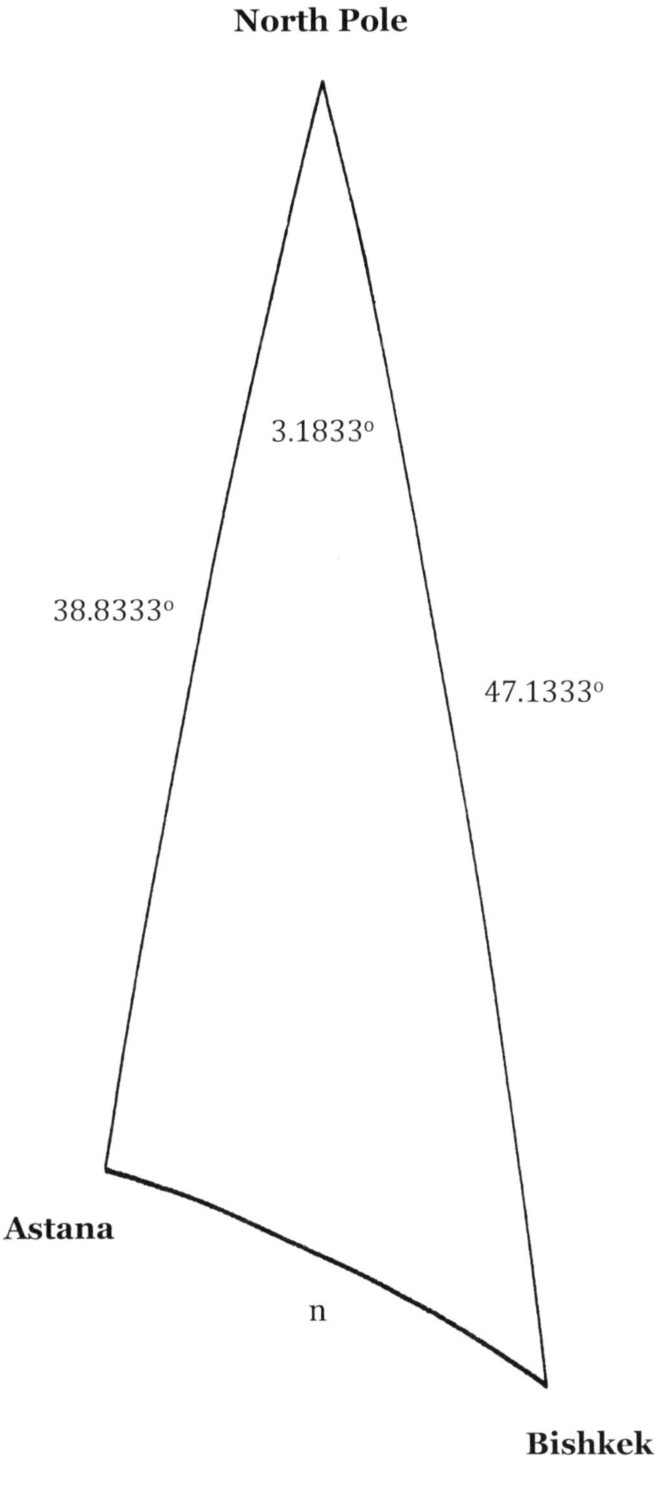
North Pole
3.1833°
38.8333°
47.1333°
Astana
n
Bishkek

Chapter 5
Problem 2
Solution

Moscow, Russia is located at $55^o 45' \text{ N} = \left(55 + {}^{45}/_{60}\right)^o \text{ N} = 55.7500^o \text{ N}$,
$37^o 36' \text{ E} = \left(37 + {}^{36}/_{60}\right)^o \text{ E} = 37.6000^o \text{ E}$.

Dushanbe, Tajikistan is located at $38^o 33' \text{ N} = \left(38 + {}^{33}/_{60}\right)^o \text{ N} = 38.5500^o \text{ N}$,
$68^o 46' \text{ E} = \left(68 + {}^{46}/_{60}\right)^o \text{ E} \approx 68.7667^o \text{ E}$.

Two sides of the spherical triangle are $90^o - 55.7500^o = 34.2500^o$ and $90^o - 38.5500^o = 51.4500^o$, respectively. The angle at the North Pole is $68.7667^o - 37.6000^o = 31.1667^o$.

See the illustration on the next page.

It follows that

$$\cos n = \cos(34.2500^o)\cos(51.4500^o) + \sin(34.2500^o)\ \sin(51.4500^o)\cos(31.1667^o)$$

$$\cos n \approx 0.891749421908 \ \Rightarrow\ n \approx \cos^{-1}(0.891749421908\ ) \ \Rightarrow\ n \approx 26.9060916617^o.$$

Therefore, the approximate distance between the two capital cities is

$$(26.9060916617^o)\left({}^{\pi}/_{180^o}\right)(3959 \text{ miles})$$

$$\approx (0.46959988834)(3959 \text{ miles})$$

$$\approx\ 1{,}859\ \text{ miles.}$$

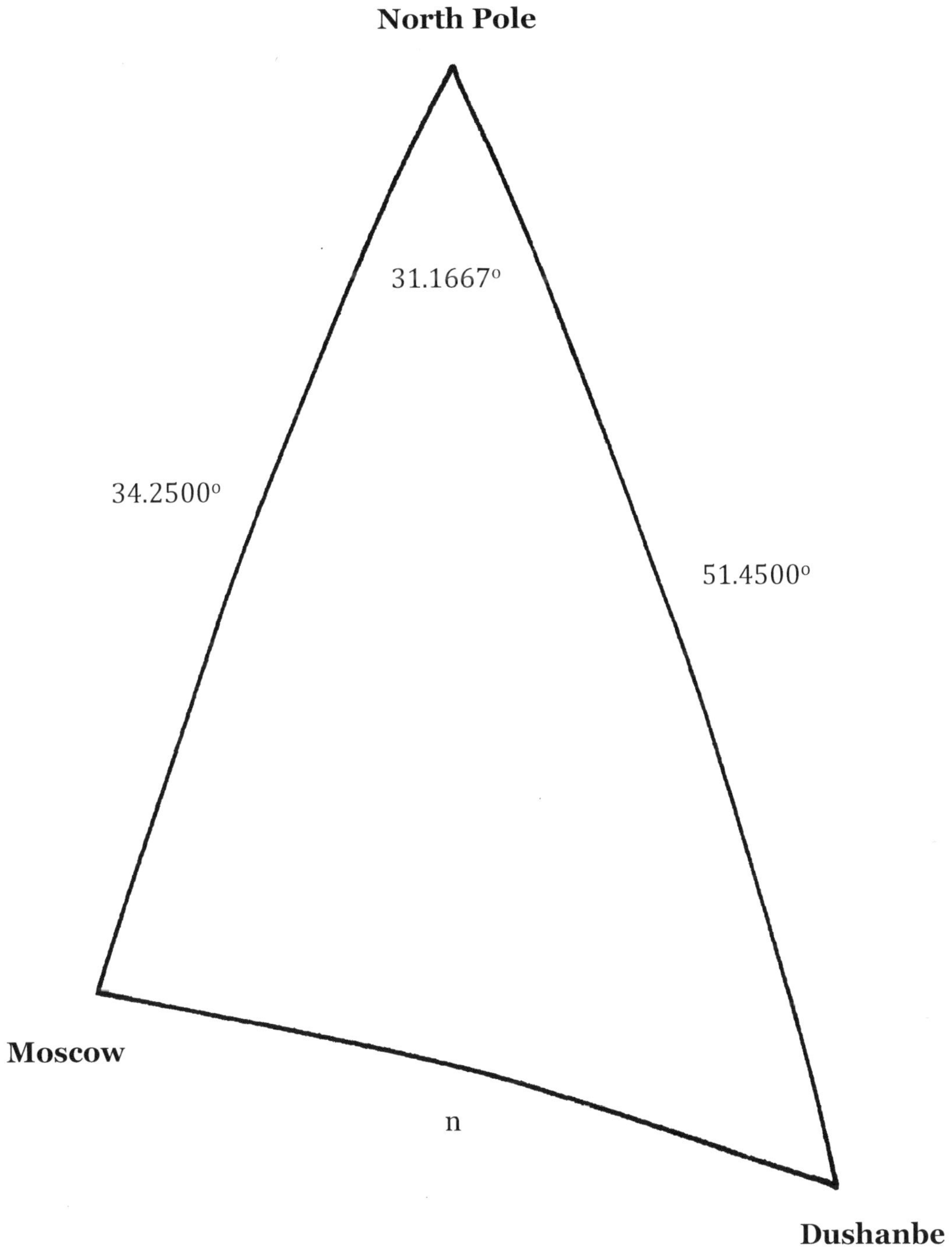
North Pole
31.1667°
34.2500°
51.4500°
Moscow
n
Dushanbe

Chapter 6

Europe

Example 1

Reykjavik, Iceland is located at $64^{o}9' \text{ N} = \left(64 + {}^{9}/_{60}\right)^{o} \text{ N} = 64.1500^{o} \text{ N}$, $21^{o}57' \text{ W} = \left(21 + {}^{57}/_{60}\right)^{o} \text{ W} = 21.9500^{o} \text{ W}$.

Stockholm, Sweden is located at $59^{o}20' \text{ N} = \left(59 + {}^{20}/_{60}\right)^{o} \text{ N} \approx 59.3333^{o} \text{ N}$, $18^{o}3' \text{ E} = \left(18 + {}^{3}/_{60}\right)^{o} \text{ E} = 18.0500^{o} \text{ E}$.

Two sides of the spherical triangle are $90^{o} - 64.1500^{o} = 25.8500^{o}$ and $90^{o} - 59.3333^{o} = 30.6667^{o}$, respectively. The angle at the North Pole is $18.0500^{o} + 21.9500^{o} = 40^{o}$.

See the illustration on the next page.

It follows that

$$\cos n = \cos(25.8500^{o})\cos(30.6667^{o}) + \sin(25.8500^{o})\ \sin(30.6667^{o})\cos(40^{o})$$

$$\cos n \approx 0.9444396923 \ \Rightarrow\ n \approx \cos^{-1}(0.9444396923) \ \Rightarrow\ n \approx 19.1889648568^{o}.$$

Therefore, the approximate distance between the two capital cities is
$(19.1889648568^{o})\left({}^{\pi}/_{180^{o}}\right)(3959 \text{ miles})$

$\approx (0.3349106168)(3959 \text{ miles})$

$\approx$ 1,326 miles.

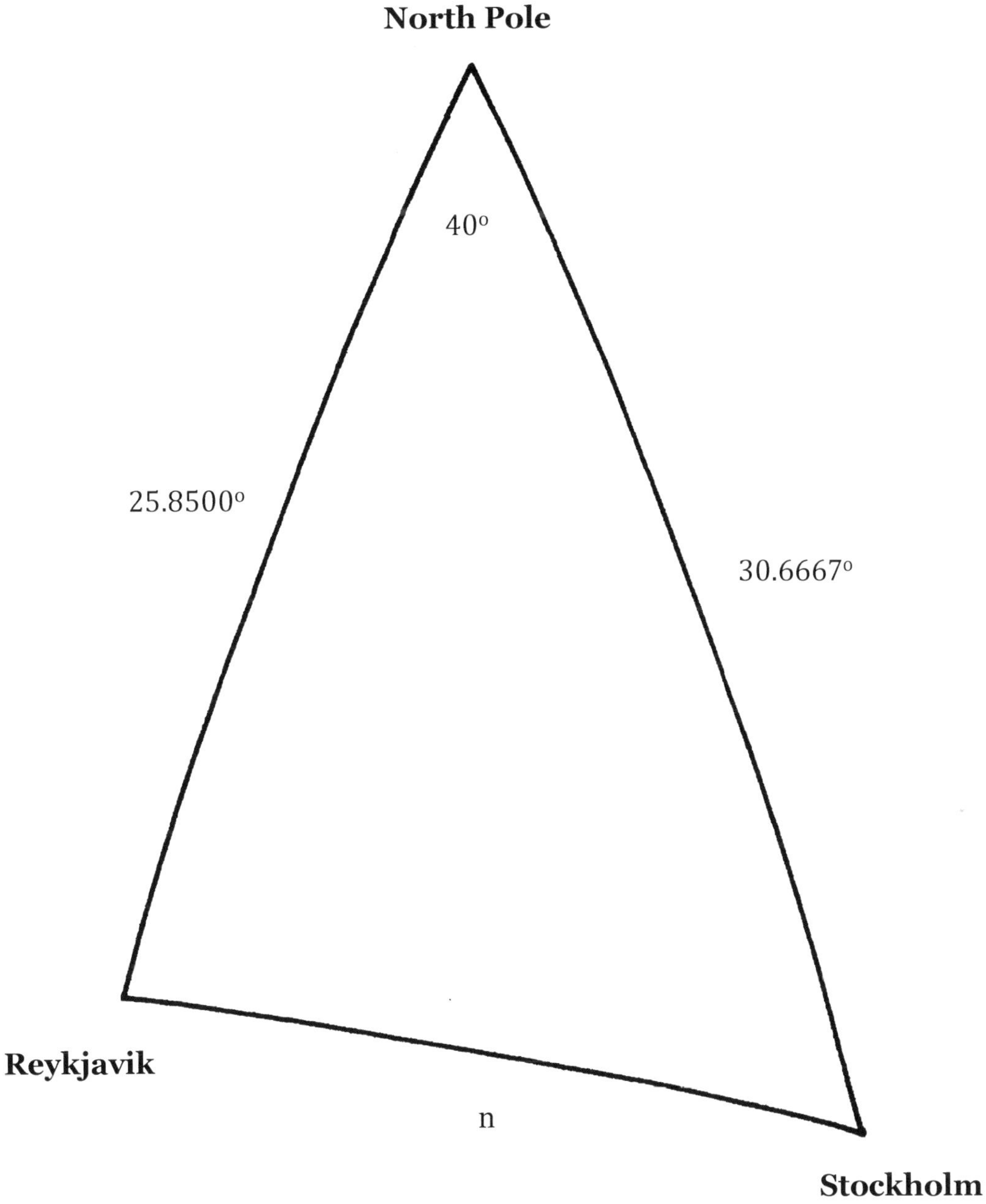
North Pole
40°
25.8500°
30.6667°
Reykjavik
n
Stockholm

Example 2

Lisbon, Portugal is located at $38°43' \text{ N} = \left(38 + {}^{43}/_{60}\right)^{\circ} \text{ N} = 38.7167° \text{ N}$,

$9°8' \text{ W} = \left(9 + {}^{8}/_{60}\right)^{\circ} \text{ W} \approx 9.1333° \text{ W}$

Riga, Latvia is located at $56°57' \text{ N} = \left(56 + {}^{57}/_{60}\right)^{\circ} \text{ N} = 56.9500° \text{ N}$,

$24°6' \text{ E} = \left(24 + {}^{6}/_{60}\right)^{\circ} \text{ E} = 24.1000° \text{ E}$

Two sides of the spherical triangle are $90° - 38.7167° = 51.2833°$ and $90° - 56.9500° = 33.0500°$, respectively. The angle at the North Pole is $24.1000° + 9.1333° = 33.2333°$.

See the illustration on the next page.

It follows that

$\cos n = \cos(51.2833°)\cos(33.0500°) + \sin(51.2833°)\ \sin(33.0500°)\cos(33.2333°)$

$\cos n \approx 0.8801941044 \Rightarrow n \approx \cos^{-1}(0.8801941044) \Rightarrow n \approx 28.33421302°$.

Therefore, the approximate distance between the two capital cities is

$(28.33421302°)\left({}^{\pi}/_{180°}\right)(3959 \text{ miles})$

$\approx (0.4945253081)(3959 \text{ miles})$

$\approx$ 1,958 miles.

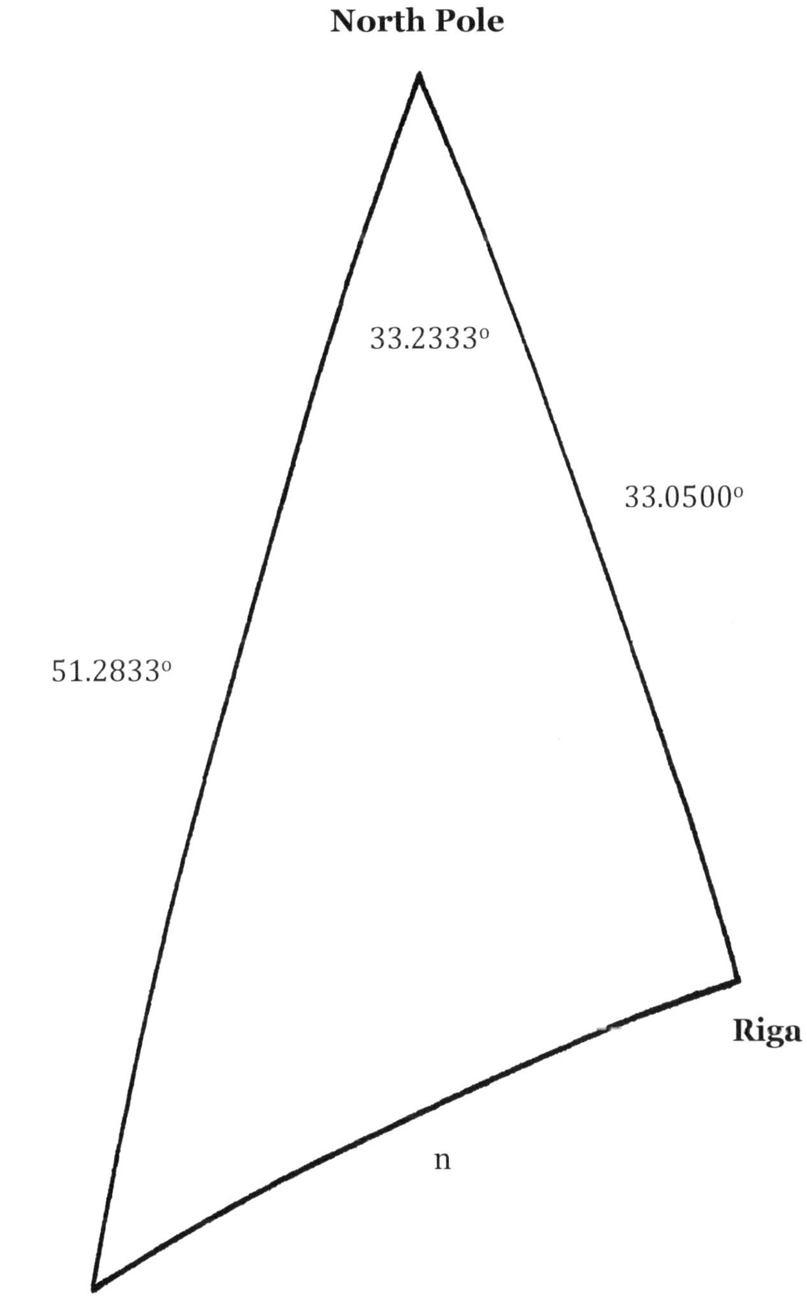
North Pole
33.2333°
33.0500°
51.2833°
Riga
n
Lisbon

Chapter 6 Problems:

Approximate the distance between the following capitals in Europe.

1. Tirana, Albania (41°19′ N 19°49′ E)
   to Skopje, Macedonia (42°0′ N 21°26′ E)

2. Andorra la Vella, Andorra (42°30′ N 1°31′ E)
   to Monaco, Monaco (43°44′ N 7°25′ E)

3. Vienna, Austria (48°12′ N 16°22′ E)
   to Podgorica, Montenegro (42°26′ N 19°16′ E)

4. Sofia, Bulgaria (42°41′ N 23°19′ E)
   to Vilnius, Lithuania (54°41′ N 25°19′ E)

5. Copenhagen, Denmark (55°40′ N 12°35′ E)
   to Minsk, Belarus (53°54′ N 27°34′ E)

6. Torshavn, Faroe Islands (62°0′ N 6°46′ W)
   to Gibraltar, Gibraltar (36°8′ N 5°21′ W)

7. Paris, France (48°52′ N 2°20′ E)
   to Madrid, Spain (40°24′ N 3°41′ W)

8. Saint Peter Port, Guernsey (49°27′ N 2°32′ W)
   to Episkopi Cantonment, Akrotiri (34°40′ N 32°51′ E)

9. Vatican City, Holy See (41°54′ N 12°27′ E)
   to Bucharest, Romania (44°26′ N 26°6′ E)

10. Budapest, Hungary (47°30′ N 19°5′ E)
    to Tallinn, Estonia (59°26′ N 24°43′ E)

11. Dublin, Ireland (53°19′ N 6°14′ W)
    to Prague, Czech Republic (50°5′ N 14°28′ E)

12. Douglas, Isle of Man (54°9′ N 4°29′ W)
    to Sarajevo, Bosnia and Herzegovina (43°52′ N 18°25′ E)

13. Rome, Italy (41°54′ N 12°29′ E)
    to Zagreb, Croatia (45°48′ N 16°0′ E)

14. Saint Helier, Jersey (49°11′ N 2°6′ W)
    to Brussels, Belgium (50°50′ N 4°20′ E)

15. Vaduz, Liechtenstein (47°8′ N 9°31′ E)
to Berlin, Germany (52°31′ N 13°24′ E)

16. Luxembourg, Luxembourg (49°36′ N 6°7′ E)
to Chisinau, Moldova (47°0′ N 28°51′ E)

17. Amsterdam, Netherlands (52°21′ N 4°55′ E)
to Ljubljana, Slovenia (46°3′ N 14°31′ E)

18. Oslo, Norway (59°55′ N 10°45′ E)
to Athens, Greece (37°59′ N 23°44′ E)

19. Warsaw, Poland (52°15′ N 21°0′ E)
to Nicosia, Cyprus (35°10′ N 33°22′ E)

20. San Marino, San Marino (43°56′ N 12°25′ E)
to Kyiv, Ukraine (50°26′ N 30°31′ E)

21. Belgrade, Serbia (44°50′ N 20°30′ E)
to Helsinki, Finland (60°10′ N 24°56′ E)

22. Bern, Switzerland (46°55′ N 7°28′ E)
to Valletta, Malta (35°53′ N 14°30′ E)

23. Longyearbyen, Svalbard (78°13′ N 15°38′ E)
to Pristina, Kosovo (42°40′ N 21°10′ E)

24. London, United Kingdom (51°30′ N 0°5′ W)
to Bratislava, Slovakia (48°9′ N 17°7′ E)

Chapter 6

Problem 1

Solution

Tirana, Albania is located at $41^o19'\ \mathrm{N} = \left(41 + {}^{19}\!/_{60}\right)^o\ \mathrm{N} \approx 41.3167^o\ \mathrm{N}$,

$19^o49'\ \mathrm{E} = \left(19 + {}^{49}\!/_{60}\right)^o\ \mathrm{E} \approx 19.8167^o\ \mathrm{E}$.

Skopje, Macedonia is located at $42^o0'\ \mathrm{N} = 42^o\ \mathrm{N}$,

$21^o26'\ \mathrm{E} = \left(21 + {}^{26}\!/_{60}\right)^o\ \mathrm{E} \approx 21.4333^o\ \mathrm{E}$.

Two sides of the spherical triangle are $90^o - 41.3167^o = 48.6833^o$ and $90^o - 42^o = 48^o$, respectively. The angle at the North Pole is $21.4333^o - 19.8167^o = 1.6166^o$.

See the illustration on the next page.

It follows that

$$\cos n = \cos(48.6833^o)\cos(48^o) + \sin(48.6833^o)\ \sin(48^o)\cos(1.6166^o)$$

$$\cos n \approx 0.999706733147 \ \Rightarrow\ n \approx \cos^{-1}(0.999706733147) \ \Rightarrow\ n \approx 1.38764934063^o.$$

Therefore, the approximate distance between the two capital cities is

$$(1.38764934063^o)\left({}^{\pi}\!/_{180^o}\right)(3959 \text{ miles})$$

$$\approx (0.024219049857)(3959 \text{ miles})$$

$$\approx 96 \text{ miles}.$$

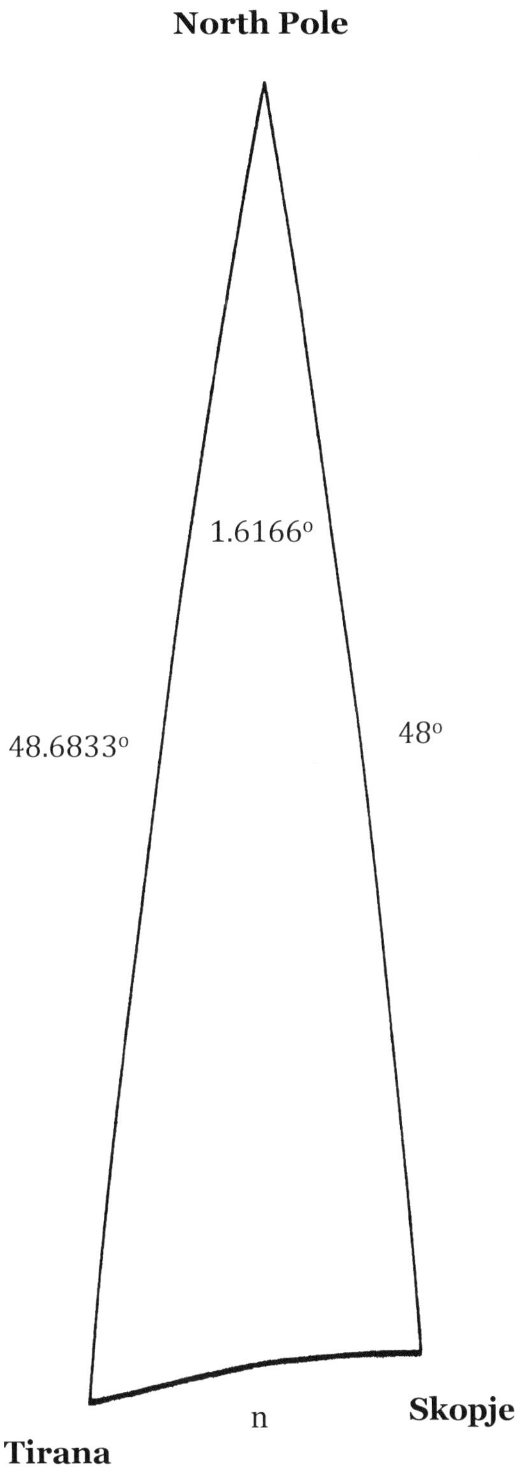
North Pole
1.6166°
48.6833°
48°
n
Skopje
Tirana

Chapter 6

Problem 2

Solution

Andorra la Vella, Andorra is located at

$42°30' \text{ N} = \left(42 + {}^{30}\!/_{60}\right)^{\circ} \text{ N} = 42.5000° \text{ N}$,

$1°31' \text{ E} = \left(1 + {}^{31}\!/_{60}\right)^{\circ} \text{ E} \approx 1.5167° \text{ E}$.

Monaco, Monaco is located at $43°44' \text{ N} = \left(43 + {}^{44}\!/_{60}\right)^{\circ} \text{ N} = 43.7333° \text{ N}$,

$7°25' \text{ E} = \left(7 + {}^{25}\!/_{60}\right)^{\circ} \text{ E} \approx 7.4167° \text{ E}$.

The sides of the spherical triangle have measures of $90° - 42.5000° = 47.5000°$ and $90° - 43.7333° = 46.2667°$, respectively. The angle opposite side n has a measure of $7.4167° - 1.5167° = 5.9000°$.

See the illustration on the next page.

It follows that

$\cos n = \cos(47.5000°)\cos(46.2667°) + \sin(47.5000°)\ \sin(46.2667°)\cos(5.9000°)$

$\cos n \approx 0.996946368627 \ \Rightarrow\ n \approx \cos^{-1}(0.996946368627) \ \Rightarrow\ n \approx 4.47874680344°$.

Therefore, the approximate distance between the two capital cities is

$(4.47874680344°)\left({}^{\pi}\!/_{180°}\right)(3959 \text{ miles})$

$\approx (0.078168878083)(3959 \text{ miles})$

$\approx\ 309\ \text{ miles.}$

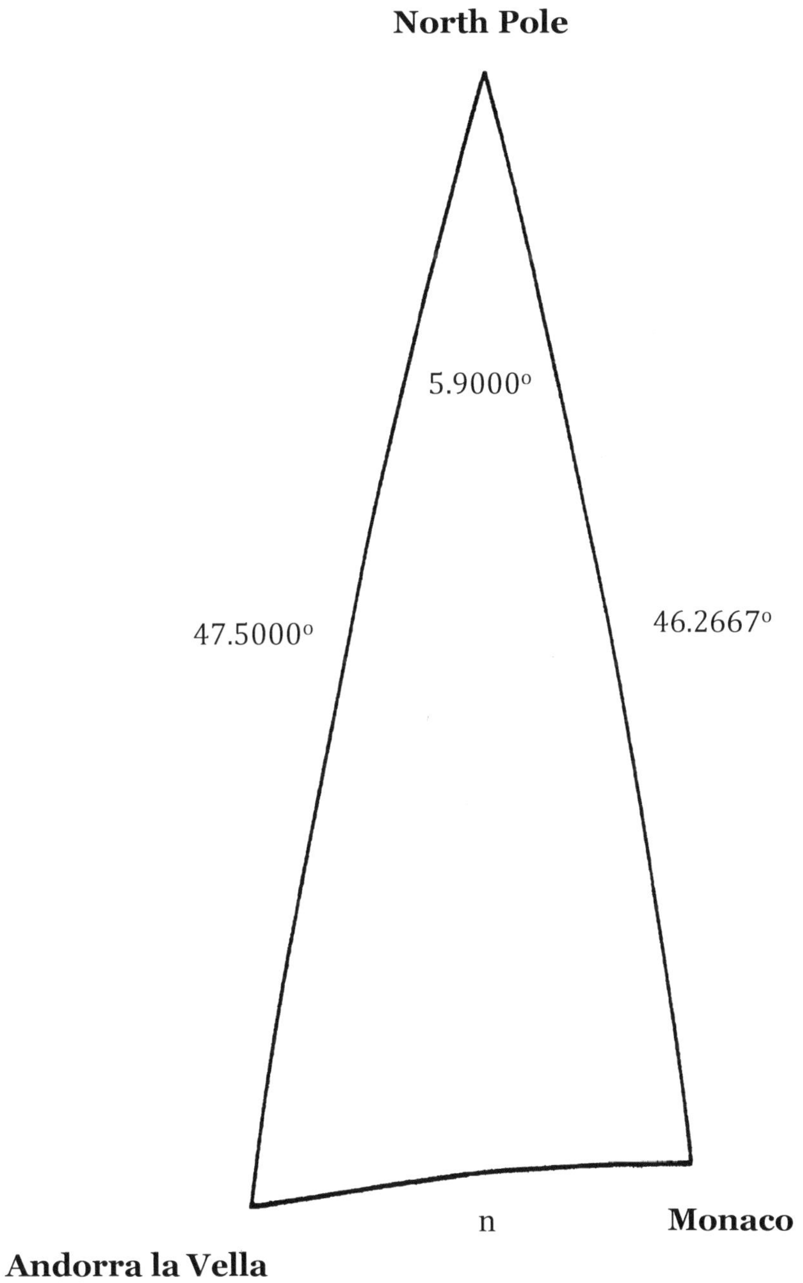
North Pole
5.9000°
47.5000°
46.2667°
n
Monaco
Andorra la Vella

Chapter 6
Problem 3
Solution

Vienna, Austria is located at $48^o12'$ N $= \left(48 + {}^{12}/_{60}\right)^o$ N $= 48.2000^o$ N,

$16^o22'$ E $= \left(16 + {}^{22}/_{60}\right)^o$ E $\approx 16.3667^o$ E.

Podgorica, Montenegro is located at $42^o26'$ N $= \left(42 + {}^{26}/_{60}\right)^o$ N $\approx 42.4333^o$ N,

$19^o16'$ E $= \left(19 + {}^{16}/_{60}\right)^o$ E $\approx 19.2667^o$ E.

Two sides of the spherical triangle are $90^o - 48.2000^o = 41.8000^o$ and $90^o - 42.4333^o = 47.5667^o$, respectively. The angle at the North Pole is $19.2667^o - 16.3667^o = 2.9000^o$.

See the illustration on the next page.

It follows that

$\cos n = \cos(41.8000^o)\cos(47.5667^o) + \sin(41.8000^o)\ \sin(47.5667^o)\cos(2.9000^o)$

$\cos n \approx 0.994309270695 \;\Rightarrow\; n \approx \cos^{-1}(0.994309270695) \;\Rightarrow\; n \approx 6.11544076157^o$.

Therefore, the approximate distance between the two capital cities is

$(6.11544076157^o)\left({}^{\pi}/_{180^o}\right)(3959 \text{ miles})$

$\approx (0.1067345765)(3959 \text{ miles})$

$\approx\ 423$ miles.

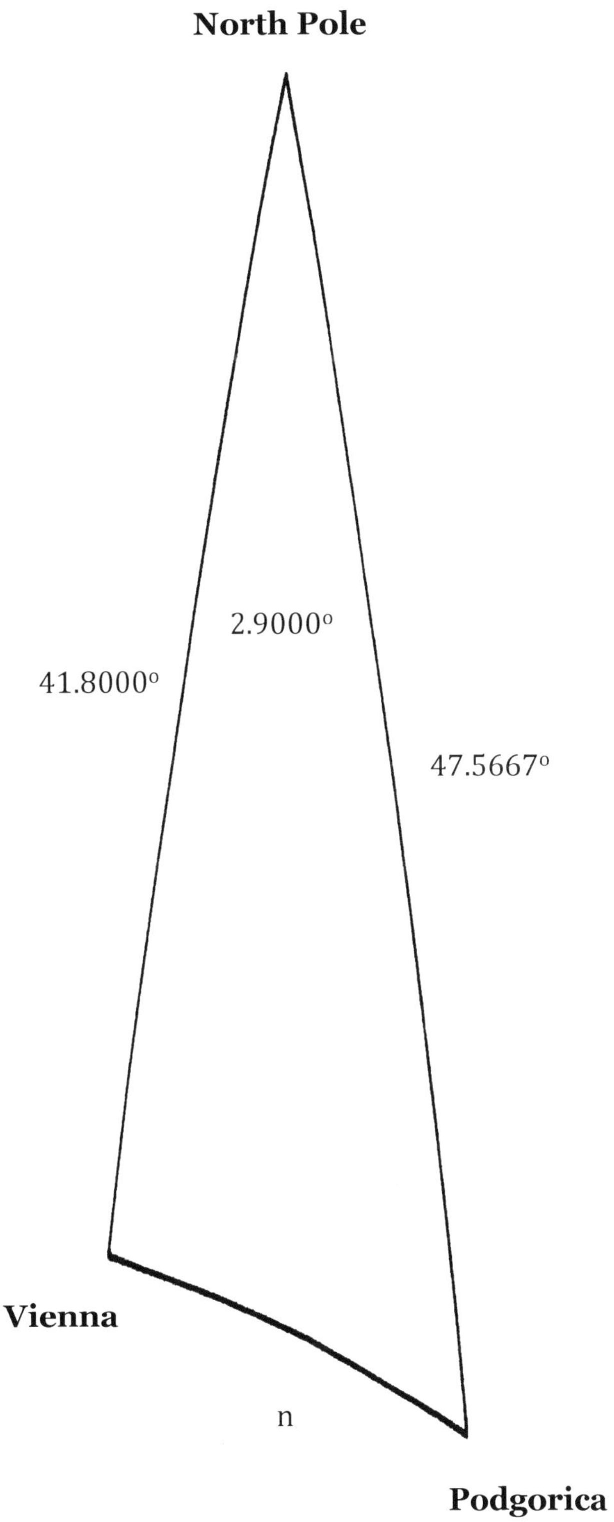

North Pole
2.9000°
41.8000°
47.5667°
Vienna
n
Podgorica

Chapter 6
Problem 4
Solution

Sofia, Bulgaria is located at $42^o41'\ \text{N} = \left(42 + {}^{41}/_{60}\right)^o\ \text{N} \approx 42.6833^o\ \text{N}$,

$23^o19'\ \text{E} = \left(23 + {}^{19}/_{60}\right)^o\ \text{E} \approx 23.3167^o\ \text{E}$.

Vilnius, Lithuania is located at $54^o41'\ \text{N} = \left(54 + {}^{41}/_{60}\right)^o\ \text{N} \approx 54.6833^o\ \text{N}$,

$25^o19'\ \text{E} = \left(25 + {}^{19}/_{60}\right)^o\ \text{E} \approx 25.3167^o\ \text{E}$.

Two sides of the spherical triangle are $90^o - 42.6833^o = 47.3167^o$ and $90^o - 54.6833^o = 35.3167^o$, respectively. The angle at the North Pole is $25.3167^o - 23.3167^o = 2^o$.

See the illustration on the next page.

It follows that

$\cos n = \cos(47.3167^o)\cos(35.3167^o) + \sin(47.3167^o)\ \sin(35.3167^o)\cos(2^o)$

$\cos n \approx 0.977888723504 \ \Rightarrow\ n \approx \cos^{-1}(0.977888723504)\ \Rightarrow\ n \approx 12.0711330148^o$.

Therefore, the approximate distance between the two capital cities is

$(12.0711330148^o)\left({}^{\pi}/_{180^o}\right)(3959\ \text{miles})$

$\approx (0.210681015554)(3959\ \text{miles})$

$\approx\ 834\ \text{miles}$.

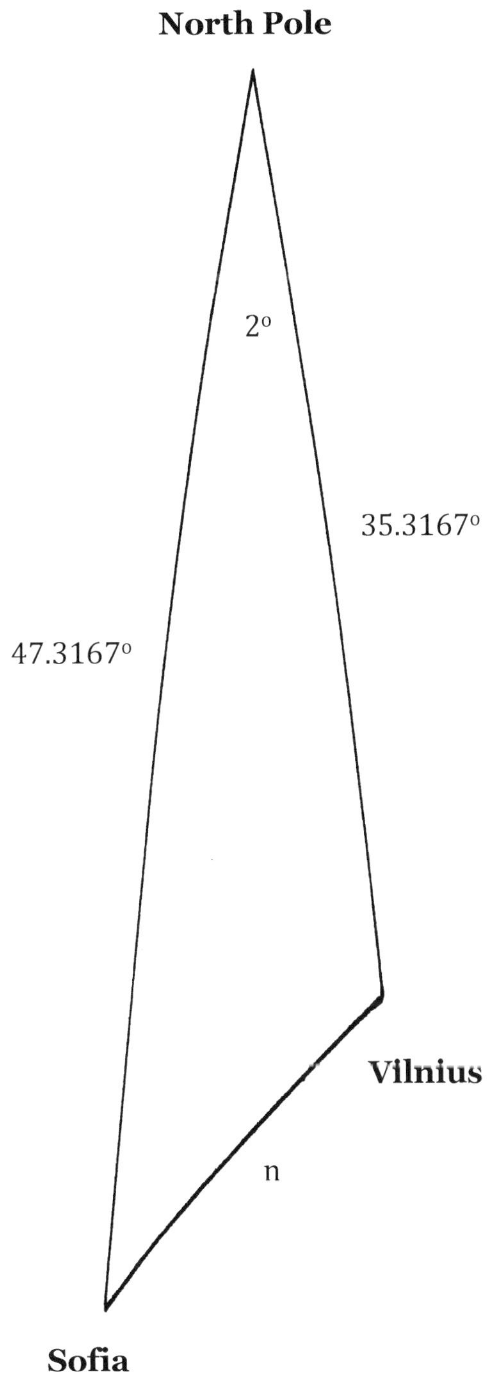
North Pole
2°
35.3167°
47.3167°
Vilnius
n
Sofia

Chapter 6

Problem 5

Solution

Copenhagen, Denmark is located at $55^o40'$ N $= \left(55 + {}^{40}/_{60}\right)^o$ N $\approx 55.6667^o$ N ,

$12^o35'$ E $= \left(12 + {}^{35}/_{60}\right)^o$ E $\approx 12.5833^o$ E .

Minsk, Belarus is located at $53^o54'$ N $= \left(53 + {}^{54}/_{60}\right)^o$ N $= 53.9000^o$ N ,

$27^o34'$ E $= \left(27 + {}^{34}/_{60}\right)^o$ E $\approx 27.5667^o$ E .

Two sides of the spherical triangle are $90^o - 55.6667^o = 34.3333^o$ and $90^o - 53.9000^o = 36.1000^o$, respectively. The angle at the North Pole is $27.5667^o - 12.5833^o = 14.9834^o$.

See the illustration on the next page.

It follows that

$\cos n = \cos(34.3333^o)\cos(36.1000^o) + \sin(34.3333^o)\ \sin(36.1000^o)\cos(14.9834^o)$

$\cos n \approx\ 0.988226352765 \Rightarrow\ n \approx \cos^{-1}(0.988226352765)\ \Rightarrow\ n \approx 8.80075974291^o$.

Therefore, the approximate distance between the two capital cities is

$(8.80075974291^o)\left({}^{\pi}/_{180^o}\right)(3959 \text{ miles})$

$\approx (0.153602234191)(3959 \text{ miles})$

$\approx\ 608\ $ miles.

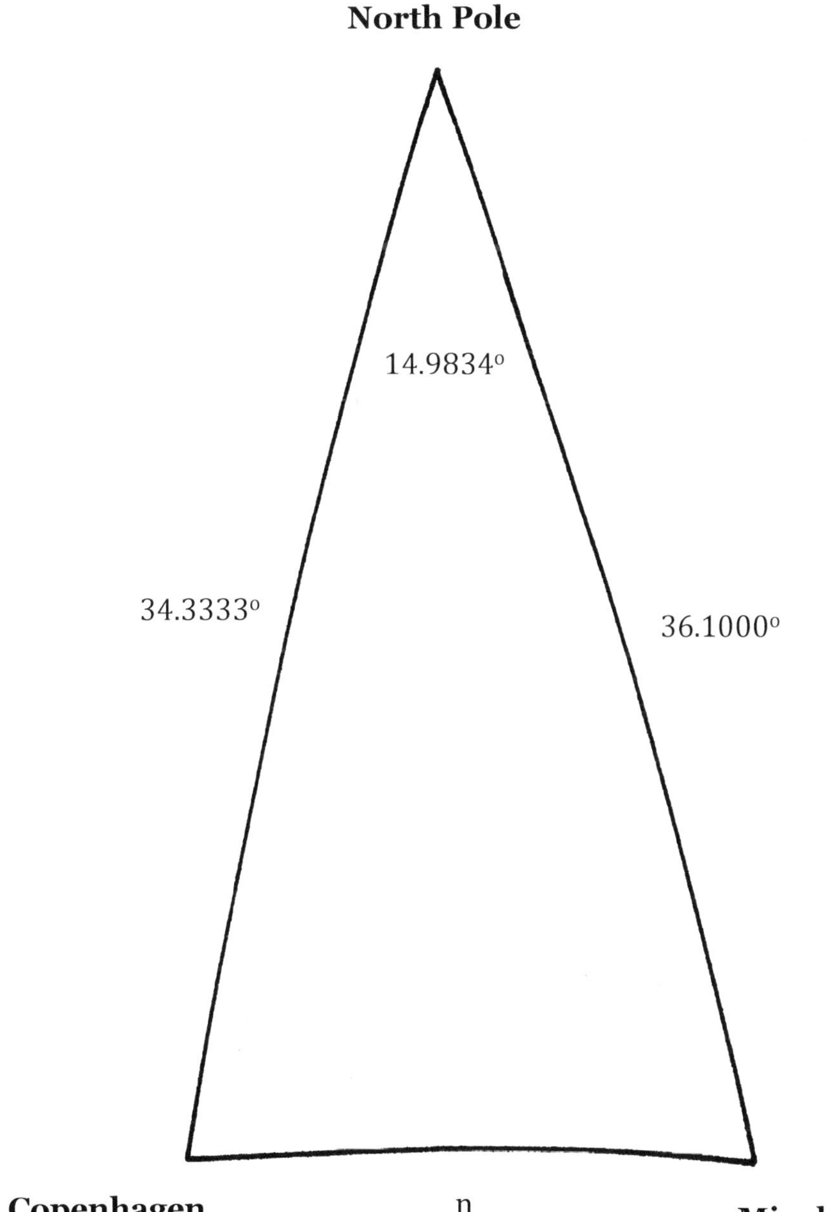
North Pole
14.9834°
34.3333°
36.1000°
Copenhagen
n
Minsk

Chapter 6
Problem 6
Solution

Torshavn, Faroe Islands is located at $62^o0' \text{ N} = 62.0000^o \text{ N}$,

$6^o46' \text{ W} = \left(6 + {}^{46}/_{60}\right)^o \text{ W} \approx 6.7667^o \text{ W}$.

Gibraltar, Gibraltar is located at $36^o8' \text{ N} = \left(36 + {}^{8}/_{60}\right)^o \text{ N} \approx 36.1333^o \text{ N}$,

$5^o21' \text{ W} = \left(5 + {}^{21}/_{60}\right)^o \text{ W} = 5.3500^o \text{ W}$.

Two sides of the spherical triangle are $90^o - 62^o = 28.0000^o$ and $90^o - 36.1333^o = 53.8667^o$, respectively. The angle at the North Pole is $6.7667^o - 5.3500^o = 1.4167^o$.

See the illustration on the next page.

It follows that

$\cos n = \cos(28.0000^o)\cos(53.8667^o) + \sin(28.0000^o)\ \sin(53.8667^o)\cos(1.4167^o)$

$\cos n \approx 0.899695592267 \ \Rightarrow\ n \approx \cos^{-1}(0.899695592267) \ \Rightarrow\ n \approx 25.8819169966^o$.

Therefore, the approximate distance between the two capital cities is

$(25.8819169966^o)\left({}^{\pi}/_{180^o}\right)(3959 \text{ miles})$

$\approx (0.451724668319)(3959 \text{ miles})$

$\approx\ 1{,}788\ \text{ miles}$.

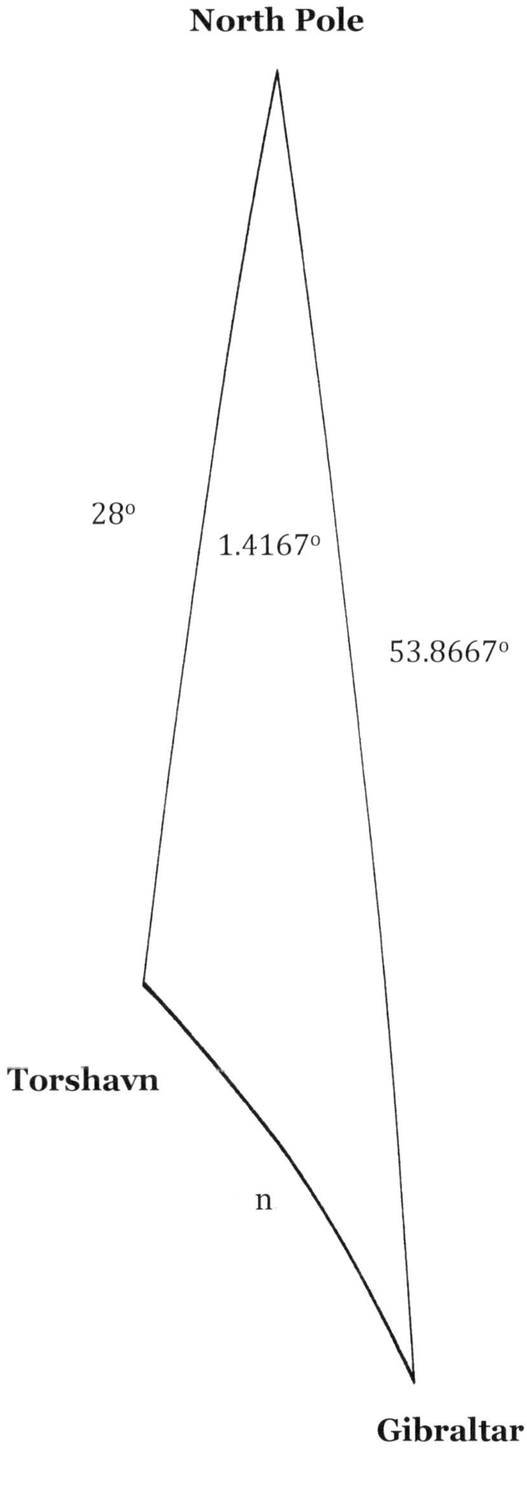
North Pole
28°
1.4167°
53.8667°
Torshavn
n
Gibraltar

Chapter 6

Problem 7

Solution

Paris, France is located at $48^o52'\text{ N} = \left(48 + {}^{52}/_{60}\right)^o \text{ N} \approx 48.8667^o\text{ N}$,

$2^o20'\text{ E} = \left(2 + {}^{20}/_{60}\right)^o \text{ E} \approx 2.3333^o\text{ E}$.

Madrid, Spain is located at $40^o24'\text{ N} = \left(40 + {}^{24}/_{60}\right)^o \text{ N} = 40.4000^o\text{ N}$,

$3^o41'\text{ W} = \left(3 + {}^{41}/_{60}\right)^o \text{ W} = 3.6833^o\text{ W}$.

Two sides of the spherical triangle are $90^o - 48.8667^o = 41.1333^o$ and $90^o - 40.4000^o = 49.6000^o$, respectively. The angle at the North Pole is $3.6833^o + 2.3333^o = 6.0166^o$.

See the illustration on the next page.

It follows that

$$\cos n = \cos(41.1333^o)\cos(49.6000^o) + \sin(41.1333^o)\ \sin(49.6000^o)\cos(6.0166^o)$$

$$\cos n \approx 0.9863421546 \ \Rightarrow n \approx \cos^{-1}(0.9863421546) \ \Rightarrow n \approx 9.480347631^o.$$

Therefore, the approximate distance between the two capital cities is

$$(9.480347631^o)\left({}^{\pi}/_{180^o}\right)(3959\text{ miles})$$

$$\approx (0.1654632804)(3959\text{ miles})$$

$$\approx \ 655\ \text{miles.}$$

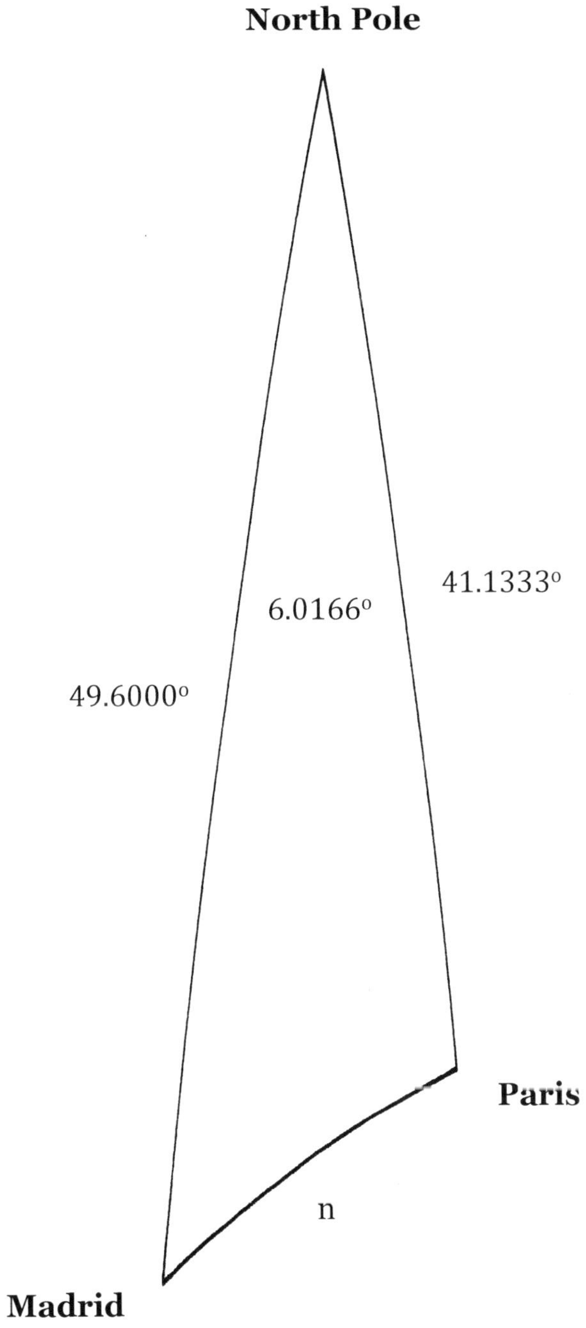
North Pole
41.1333°
6.0166°
49.6000°
Paris
n
Madrid

Chapter 6
Problem 8
Solution

Saint Peter Port, Guernsey is located at

$49^{o}27'\ \mathrm{N} = \left(49 + {}^{27}/_{60}\right)^{o}\ \mathrm{N} = 49.4500^{o}\ \mathrm{N}\ , 2^{o}32'\ \mathrm{W} = \left(2 + {}^{32}/_{60}\right)^{o}\ \mathrm{W} \approx 2.5333^{o}\ \mathrm{W}$.

Episkopi Cantonment, Akrotiri is located at

$34^{o}40'\ \mathrm{N} = \left(34 + {}^{40}/_{60}\right)^{o}\ \mathrm{N} \approx 34.6667^{o}\ \mathrm{N}\ , 32^{o}51'\ \mathrm{E} = \left(32 + {}^{51}/_{60}\right)^{o}\ \mathrm{E} = 32.8500^{o}\ \mathrm{E}$.

Two sides of the spherical triangle are $90^{o} - 49.4500^{o} = 40.5500^{o}$ and $90^{o} - 34.6667^{o} = 55.3333^{o}$, respectively. The angle at the North Pole is $2.5333^{o} + 32.8500^{o} = 35.3833^{o}$.

See the illustration on the next page.

It follows that

$\cos n = \cos(40.5500^{o})\cos(55.3333^{o}) + \sin(40.5500^{o})\ \sin(55.3333^{o})\cos(35.3833^{o})$

$\cos n \approx 0.868136858814 \Rightarrow n \approx \cos^{-1}(0.868136858814) \Rightarrow n \approx 29.7571529484^{o}$.

Therefore, the approximate distance between the two capital cities is

$(29.7571529484^{o})\left({}^{\pi}/_{180^{o}}\right)(3959\ \text{miles})$

$\approx (0.51936029497)(3959\ \text{miles})$

$\approx\ 2{,}056\ \text{miles}$.

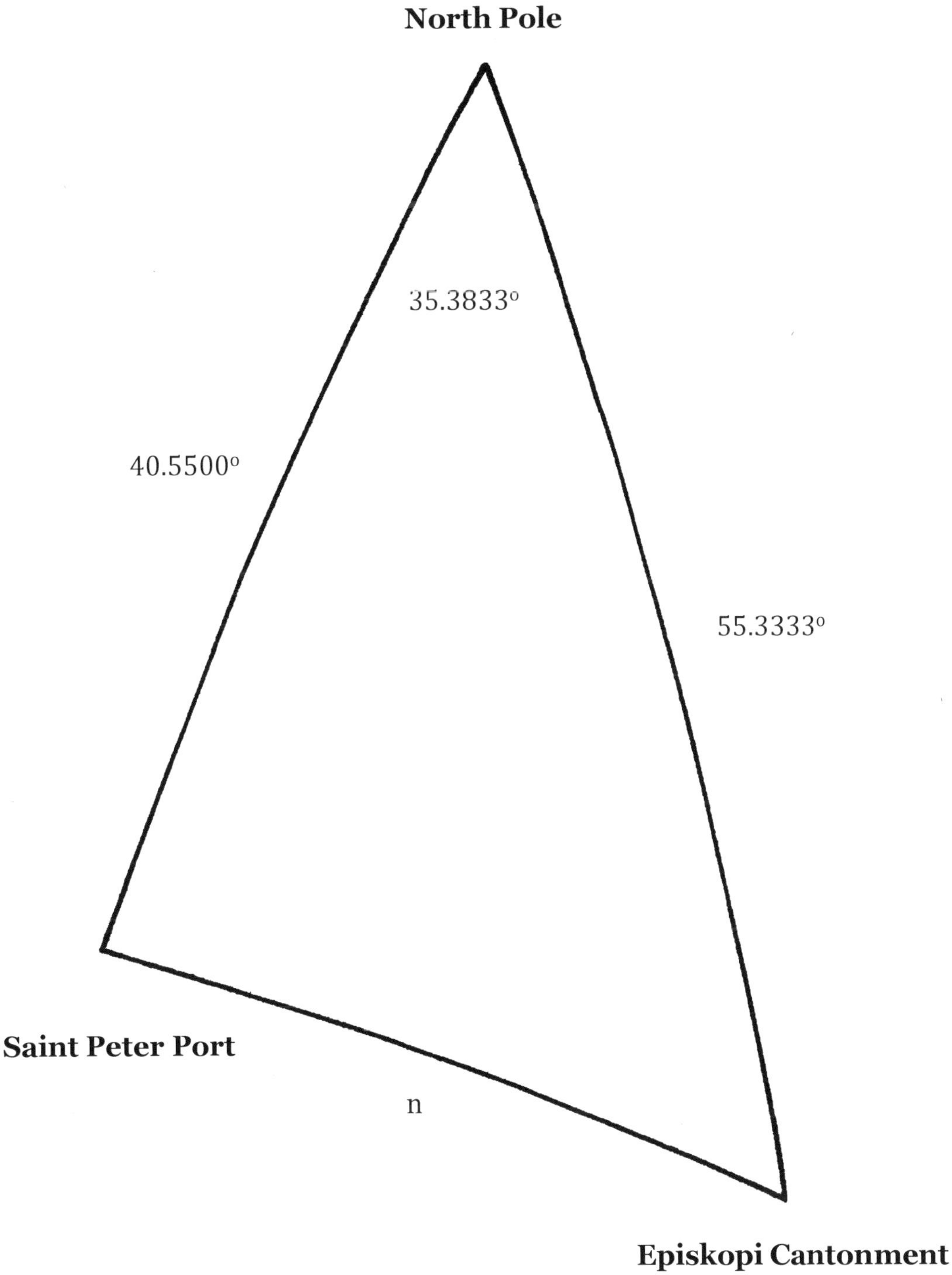
North Pole
35.3833°
40.5500°
55.3333°
Saint Peter Port
n
Episkopi Cantonment

Chapter 6

Problem 9

Solution

Vatican City, Holy See is located at $41^{o}54' \text{ N} = \left(41 + {}^{54}/_{60}\right)^{o} \text{ N} \approx 41.9000^{o} \text{ N}$, $12^{o}27' \text{ E} = \left(12 + {}^{27}/_{60}\right)^{o} \text{ E} = 12.4500^{o} \text{ E}$.

Bucharest, Romania is located at $44^{o}26' \text{ N} = \left(44 + {}^{26}/_{60}\right)^{o} \text{ N} \approx 44.4333^{o} \text{ N}$,

$26^{o}6' \text{ E} = \left(26 + {}^{6}/_{60}\right)^{o} \text{ E} = 26.1000^{o} \text{ E}$.

Two sides of the spherical triangle are $90^{o} - 41.9000^{o} = 48.1000^{o}$ and $90^{o} - 44.4333^{o} = 45.5667^{o}$, respectively. The angle at the North Pole is $26.1000^{o} - 12.4500^{o} = 13.6500^{o}$.

See the illustration on the next page.

It follows that

$\cos n = \cos(48.1000^{o}) \cos(45.5667^{o}) + \sin(48.1000^{o}) \sin(45.5667^{o}) \cos(13.6500^{o})$

$\cos n \approx 0.984011065539 \Rightarrow n \approx \cos^{-1}(0.984011065539) \Rightarrow n \approx 10.2595368093^{o}$.

Therefore, the approximate distance between the two capital cities is

$(10.2595368093^{o})\left({}^{\pi}/_{180^{o}}\right)(3959 \text{ miles})$

$\approx (0.179062697051)(3959 \text{ miles})$

≈ 709 miles.

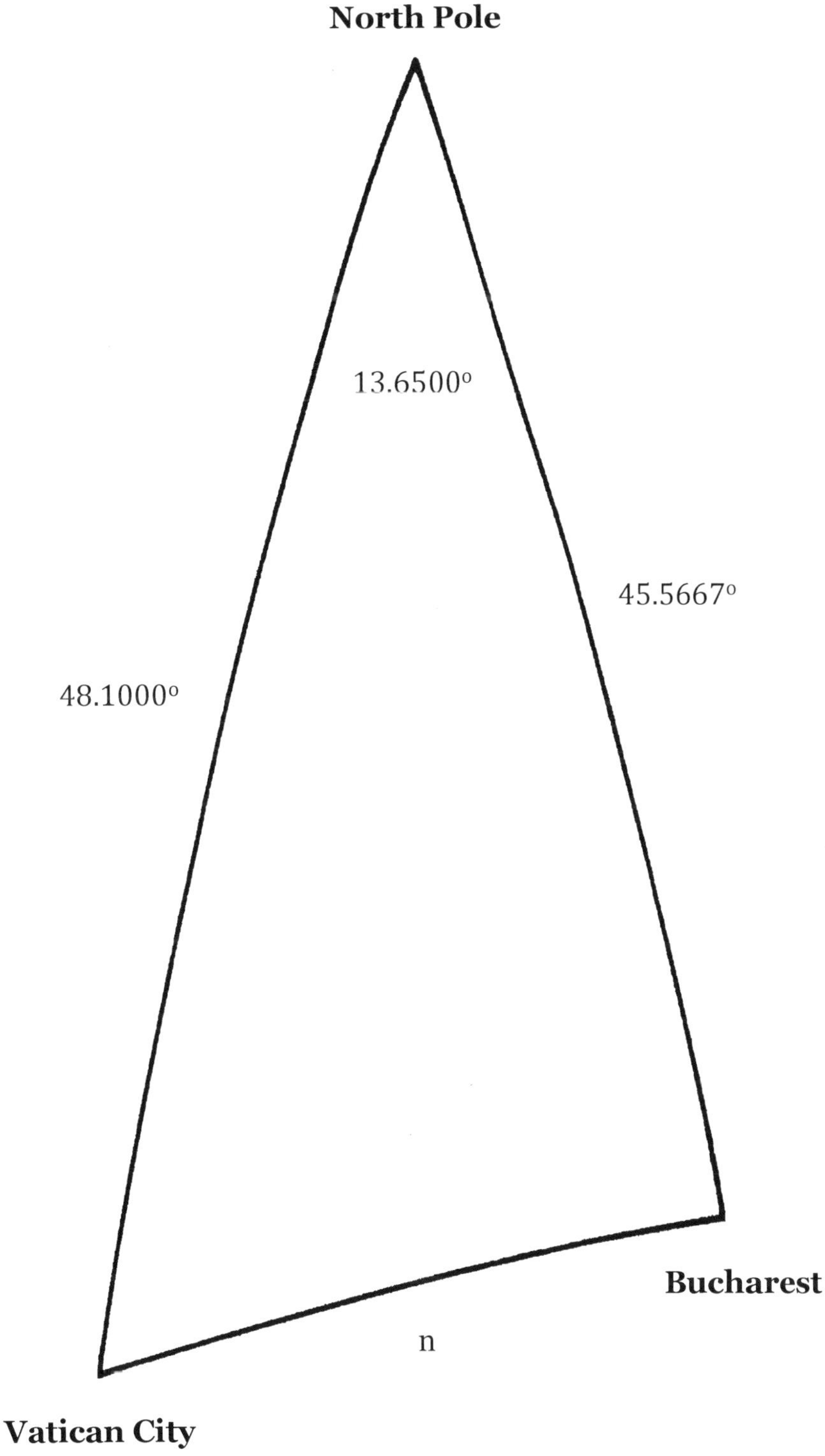
North Pole
13.6500°
45.5667°
48.1000°
Bucharest
n
Vatican City

Chapter 6

Problem 10

Solution

Budapest, Hungary is located at $47^o30'\ \text{N} = \left(47 + {}^{30}/_{60}\right)^o \text{N} = 47.5000^o\ \text{N}$,

$19^o5'\ \text{E} = \left(19 + {}^{5}/_{60}\right)^o\ \text{E} \approx 19.0833^o\ \text{E}$.

Tallinn, Estonia is located at $59^o26'\ \text{N} = \left(59 + {}^{26}/_{60}\right)^o\ \text{N} \approx 59.4333^o\ \text{N}$,

$24^o43'\ \text{E} = \left(24 + {}^{43}/_{60}\right)^o\ \text{E} \approx 24.7167^o\ \text{E}$.

Two sides of the spherical triangle have measures of $90^o - 47.5000^o = 42.5000^o$ and $90^o - 59.4333^o = 30.5667^o$, respectively. The angle opposite side n has a measure of $24.7167^o - 19.0833^o = 5.6334^o$.

See the illustration on the next page.

It follows that

$\cos n = \cos(42.5000^o)\cos(30.5667^o) + \sin(42.5000^o)\ \sin(30.5667^o)\cos(5.6334^o)$

$\cos n \approx 0.976729673194 \Rightarrow n \approx \cos^{-1}(0.976729673194) \Rightarrow n \approx 12.3846765568^o$.

Therefore, the approximate distance between the two capital cities is

$(12.3846765568^o)\left({}^{\pi}/_{180^o}\right)(3959\ \text{miles})$

$\approx (0.216153382711)(3959\ \text{miles})$

≈ 856 miles.

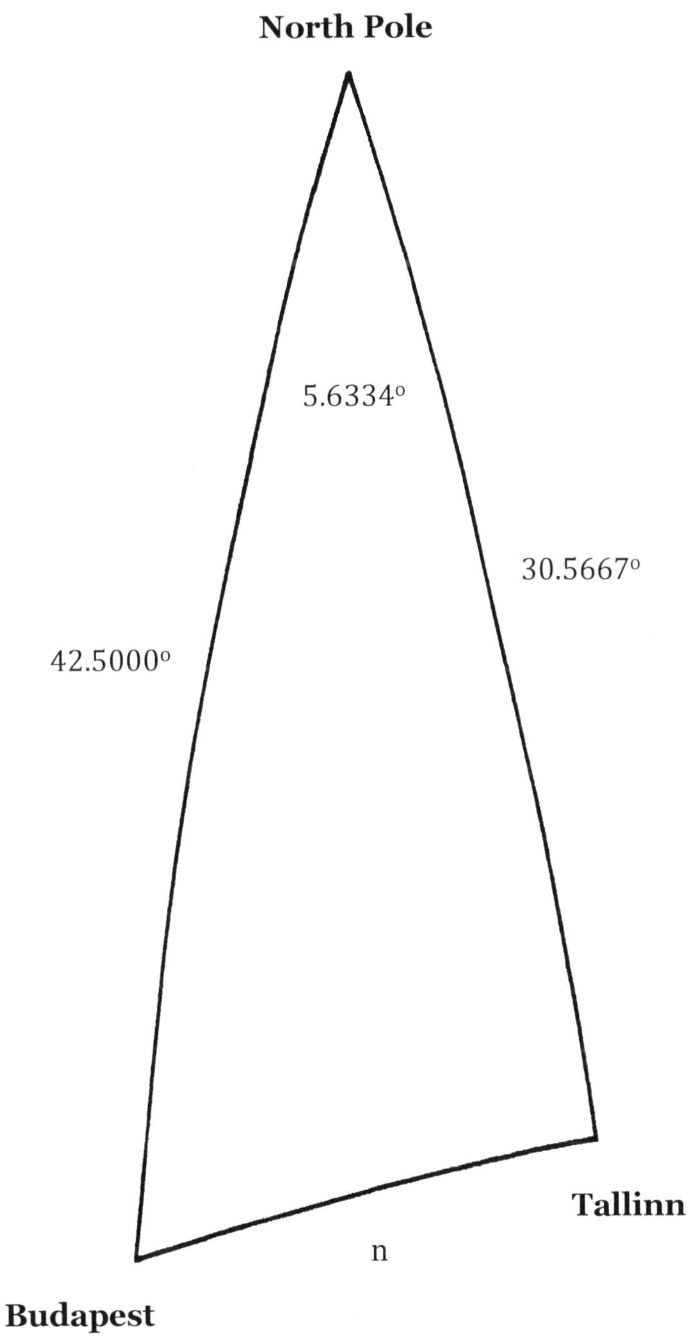
North Pole
5.6334°
30.5667°
42.5000°
Tallinn
n
Budapest

Chapter 6
Problem 11
Solution

Dublin, Ireland is located at $53^o19' \text{ N} = \left(53 + {}^{19}/_{60}\right)^o \text{ N} \approx 53.3167^o \text{ N}$,
$6^o14' \text{ W} = \left(6 + {}^{14}/_{60}\right)^o \text{ W} \approx 6.2333^o \text{ W}$

Prague, Czech Republic is located at $50^o5' \text{ N} = \left(50 + {}^{5}/_{60}\right)^o \text{ N} \approx 50.0833^o \text{ N}$,
$14^o28' \text{ E} = \left(14 + {}^{28}/_{60}\right)^o \text{ E} \approx 14.4667^o \text{ E}$

Two sides of the spherical triangle are $90^o - 53.3167^o = 36.6833^o$ and $90^o - 50.0833^o = 39.9167^o$, respectively. The angle at the North Pole is $6.2333^o + 14.4667^o = 20.7000^o$.

See the illustration on the next page.

It follows that

$$\cos n = \cos(36.6833^o)\cos(39.9167^o) + \sin(36.6833^o)\ \sin(39.9167^o)\cos(20.7000^o)$$

$$\cos n \approx 0.973661809434 \Rightarrow n \approx \cos^{-1}(0.973661809434\ ) \Rightarrow n \approx 13.1791816937^o.$$

Therefore, the approximate distance between the two capital cities is

$$(13.1791816937^o)\left({}^{\pi}/_{180^o}\right)(3959 \text{ miles})$$

$$\approx (0.230020113273)(3959 \text{ miles})$$

$$\approx 911 \text{ miles.}$$

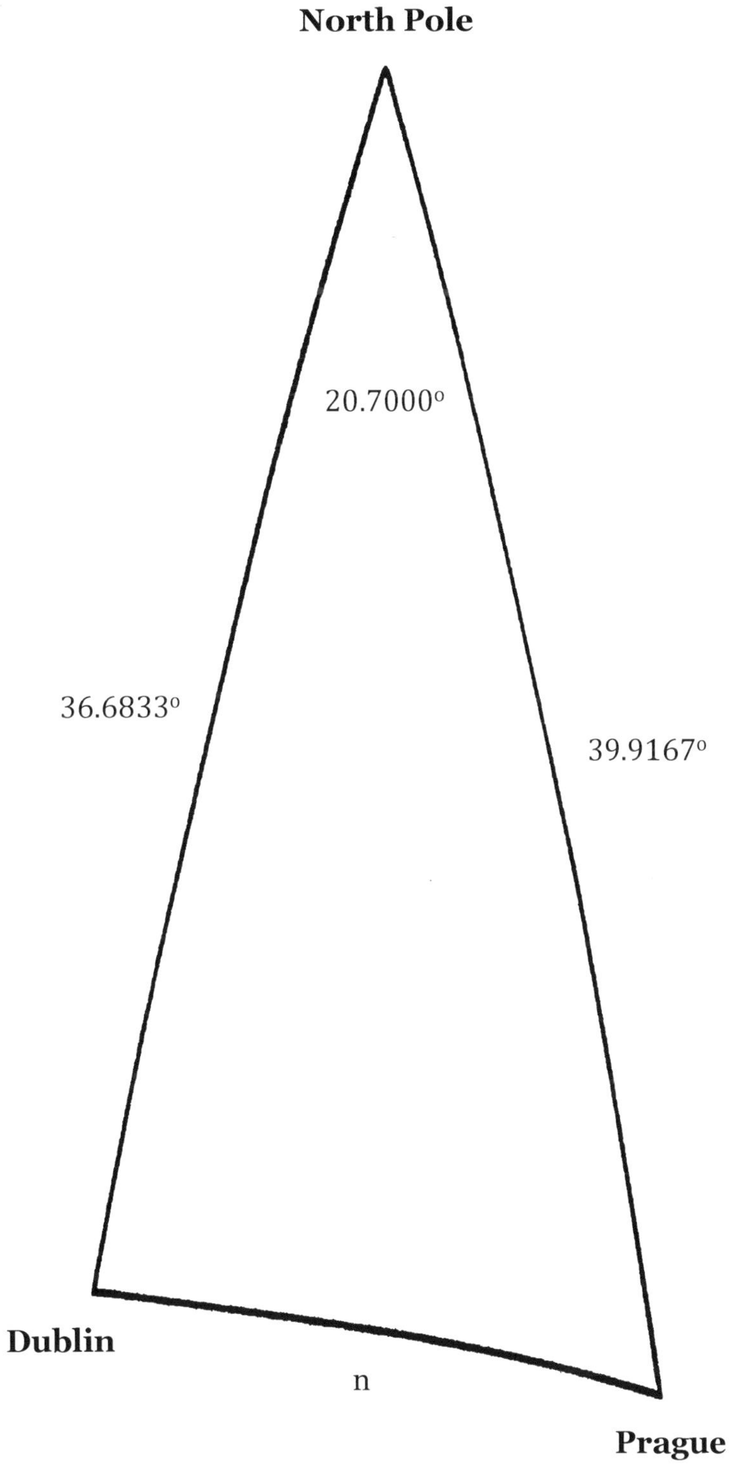
North Pole
20.7000°
36.6833°
39.9167°
Dublin
n
Prague

Chapter 6
Problem 12
Solution

Douglas, Isle of Man is located at $54^o 9' \text{ N} = \left(54 + {}^{9}/_{60}\right)^o \text{ N} = 54.1500^o \text{ N}$,

$4^o 29' \text{ W} = \left(4 + {}^{29}/_{60}\right)^o \text{ W} \approx 4.4833^o \text{ W}$.

Sarajevo, Bosnia and Herzegovina is located at

$43^o 52' \text{ N} = \left(43 + {}^{52}/_{60}\right)^o \text{ N} \approx 43.8667^o \text{ N}, 18^o 25' \text{ E} = \left(18 + {}^{25}/_{60}\right)^o \text{ E} \approx 18.4167^o \text{ E}$.

Two sides of the spherical triangle are $90^o - 54.1500^o = 35.8500^o$ and $90^o - 43.8667^o = 46.1333^o$, respectively. The angle at the North Pole is $4.4833^o + 18.4167^o = 22.9000^o$.

See the illustration on the next page.

It follows that

$\cos \text{n} = \cos(35.8500^o)\cos(46.1333^o) + \sin(35.8500^o)\ \sin(46.1333^o)\cos(22.9000^o)$

$\cos \text{n} \approx 0.950658619202 \Rightarrow \text{n} \approx \cos^{-1}(0.950658619202) \Rightarrow \text{n} \approx 18.0736297233^o$.

Therefore, the approximate distance between the two capital cities is

$(18.0736297233^o)\left({}^{\pi}/_{180^o}\right)(3959 \text{ miles})$

$\approx (0.315444346459)(3959 \text{ miles})$

$\approx 1{,}249$ miles.

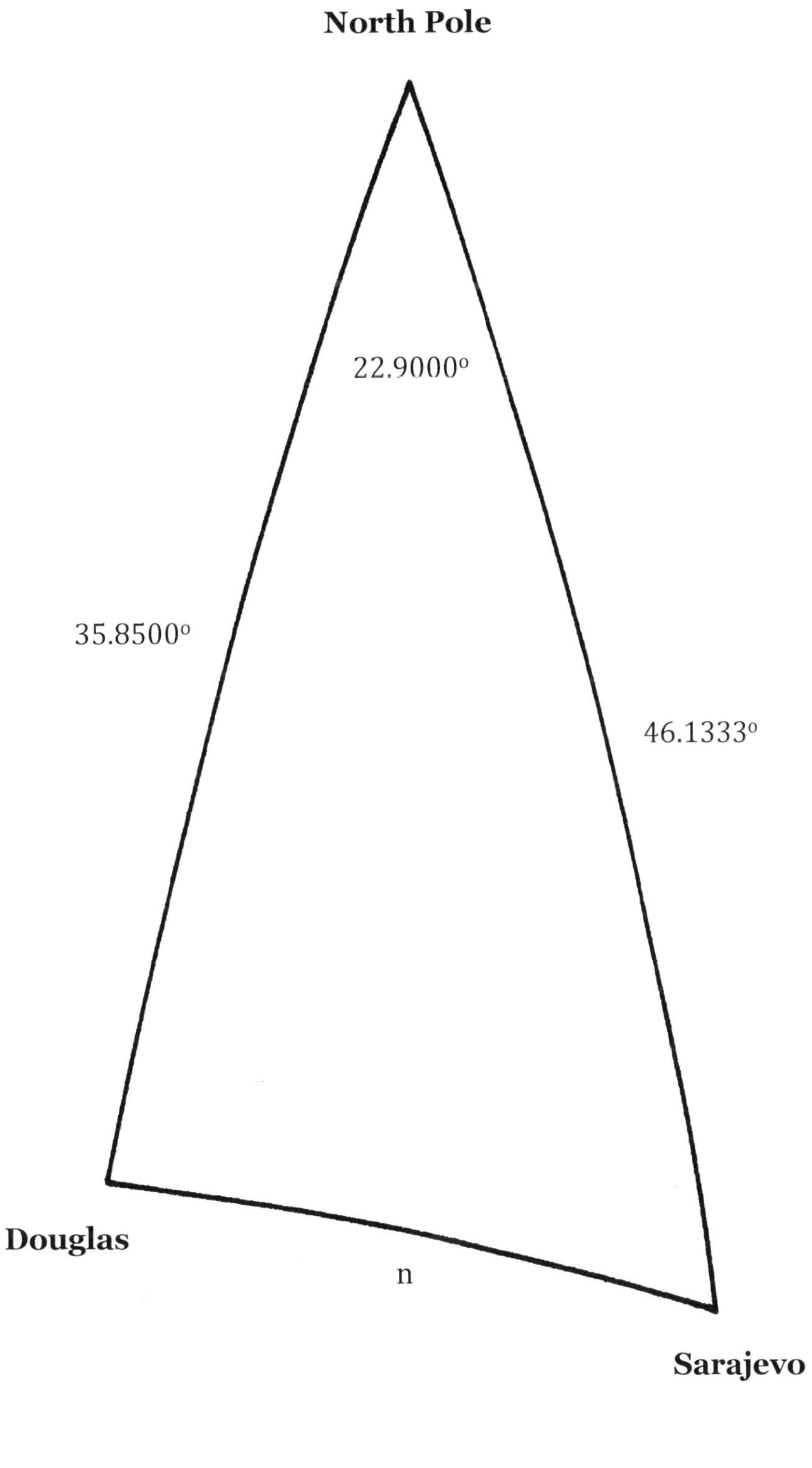

North Pole
22.9000°
35.8500°
46.1333°
Douglas
n
Sarajevo

Chapter 6

Problem 13

Solution

Rome, Italy is located at $41^{o}54'\ \text{N} = \left(41 + {}^{54}/_{60}\right)^{o}\ \text{N} = 41.9000^{o}\ \text{N}$, $12^{o}29'\ \text{E} = \left(12 + {}^{29}/_{60}\right)^{o}\ \text{E} \approx 12.4833^{o}\ \text{E}$.

Zagreb, Croatia is located at $45^{o}48'\ \text{N} = \left(45 + {}^{48}/_{60}\right)^{o}\ \text{N} = 45.8000^{o}\ \text{N}$, $16^{o}0'\ \text{E} = 16.0000^{o}\ \text{E}$.

Two sides of the spherical triangle are $90^{o} - 41.9000^{o} = 48.1000^{o}$ and $90^{o} - 45.8000^{o} = 44.2000^{o}$, respectively. The angle at the North Pole is $16.0000^{o} - 12.4833^{o} = 3.5167^{o}$.

See the illustration on the next page.

It follows that

$\cos n = \cos(48.1000^{o})\cos(44.2000^{o}) + \sin(48.1000^{o})\ \sin(44.2000^{o})\cos(3.5167^{o})$

$\cos n \approx\ 0.996707155393 \Rightarrow\ n \approx \cos^{-1}(0.996707155393)\ \Rightarrow\ n \approx 4.65095880048^{o}$.

Therefore, the approximate distance between the two capital cities is

$(4.65095880048^{o})\left({}^{\pi}/_{180^{o}}\right)(3959\ \text{miles})$

$\approx (0.081174544443)(3959\ \text{miles})$

$\approx\ 321\ \text{miles}$.

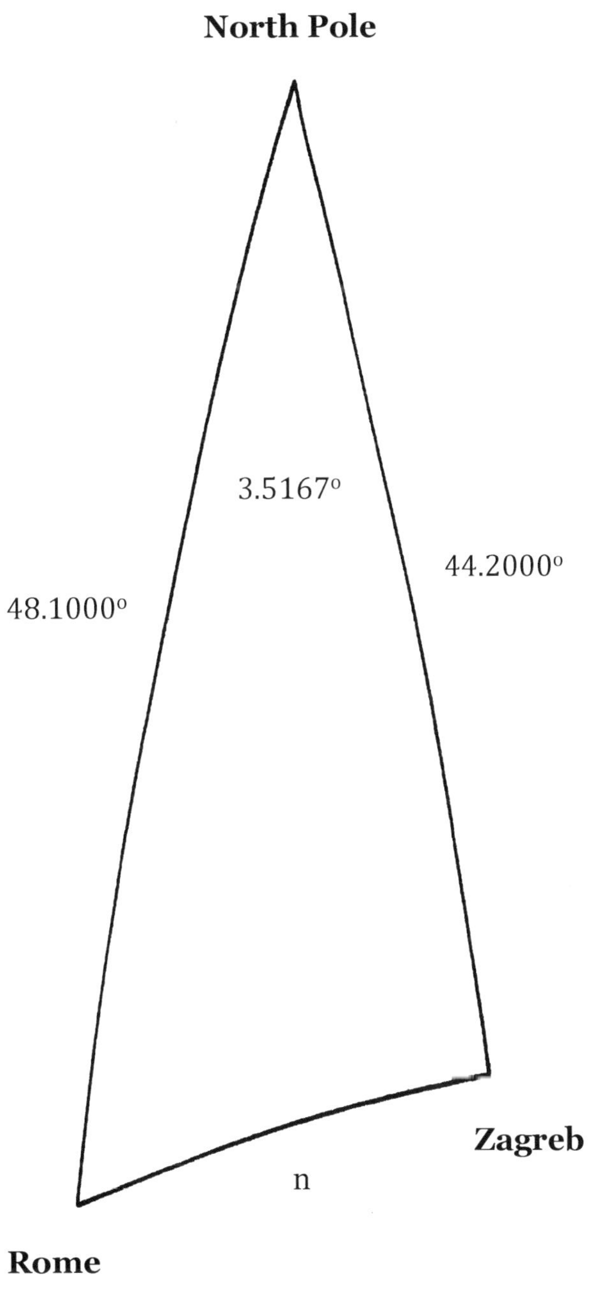
North Pole
3.5167°
48.1000°
44.2000°
Zagreb
n
Rome

Chapter 6
Problem 14
Solution

Saint Helier, Jersey is located at $49^o 11' \text{ N} = \left(49 + {}^{11}/_{60}\right)^o \text{ N} \approx 49.1833^o \text{ N}$, $2^o 6' \text{ W} = \left(2 + {}^{6}/_{60}\right)^o \text{ W} = 2.1000^o \text{ W}$.

Brussels, Belgium is located at $50^o 50' \text{ N} = \left(50 + {}^{50}/_{60}\right)^o \text{ N} = 50.8333^o \text{ N}$, $4^o 20' \text{ E} = \left(4 + {}^{20}/_{60}\right)^o \text{ E} \approx 4.3333^o \text{ E}$.

Two sides of the spherical triangle are $90^o - 49.1833^o = 40.8167^o$ and $90^o - 50.8333^o = 39.1667^o$, respectively. The angle at the North Pole is $2.1000^o + 4.3333^o = 6.4333^o$.

See the illustration on the next page.

It follows that

$\cos n = \cos(40.8167^o)\cos(39.1667^o) + \sin(40.8167^o)\ \sin(39.1667^o)\cos(6.4333^o)$

$\cos n \approx 0.996985789303 \ \Rightarrow \ n \approx \cos^{-1}(0.996985789303) \ \Rightarrow \ n \approx 4.44972920214^o$.

Therefore, the approximate distance between the two capital cities is

$(4.44972920214^o)\left({}^{\pi}/_{180^o}\right)(3959 \text{ miles})$

$\approx (0.0776624254)(3959 \text{ miles})$

$\approx \ 307 \ \text{ miles}$.

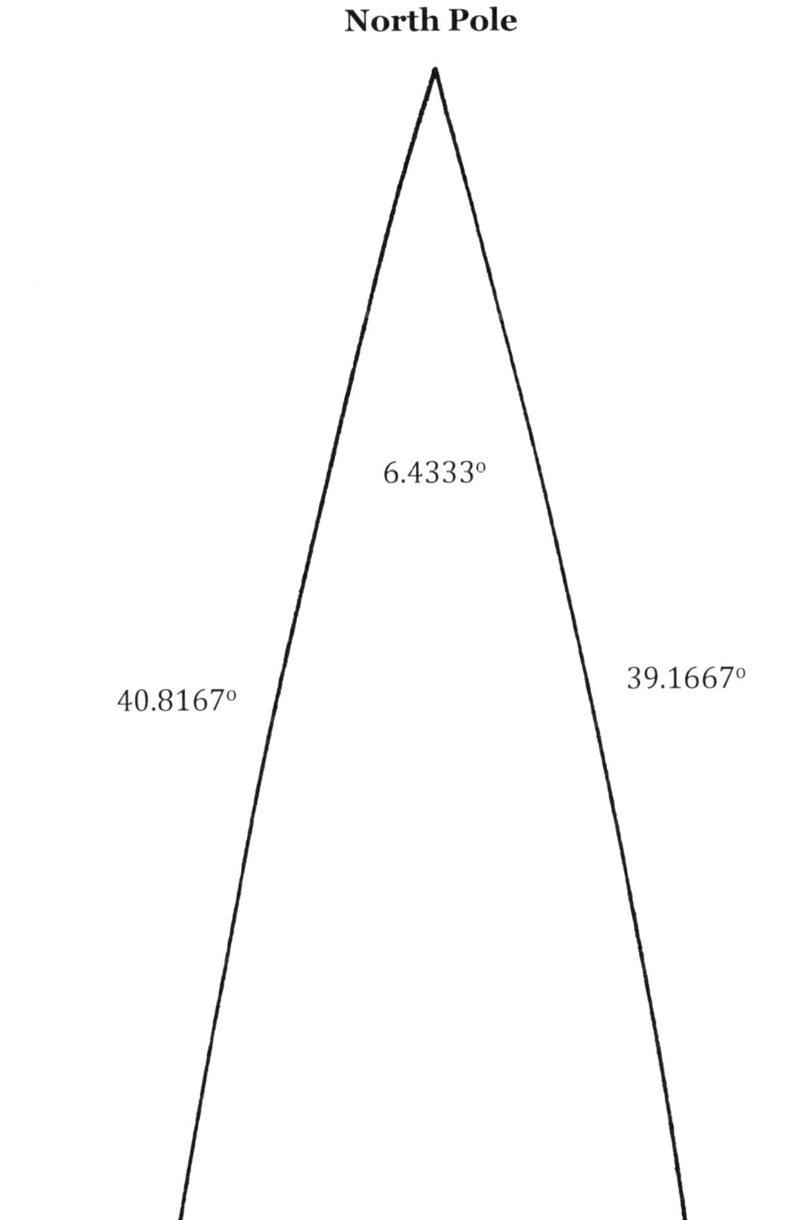
North Pole
6.4333°
40.8167°
39.1667°
Saint Helier
n
Brussels

Chapter 6

Problem 15

Solution

Vaduz, Liechtenstein is located at $47^o8' \text{ N} = \left(47 + {}^{8}/_{60}\right)^o \text{ N} \approx 47.1333^o \text{ N}$, $9^o31' \text{ E} = \left(9 + {}^{31}/_{60}\right)^o \text{ E} \approx 9.5167^o \text{ E}$.

Berlin, Germany is located at $52^o31' \text{ N} = \left(52 + {}^{31}/_{60}\right)^o \text{ N} \approx 52.5167^o \text{ N}$, $13^o24' \text{ E} = \left(13 + {}^{24}/_{60}\right)^o \text{ E} = 13.4000^o \text{ E}$.

Two sides of the spherical triangle are $90^o - 47.1333^o = 42.8667^o$ and $90^o - 52.5167^o = 37.4833^o$, respectively. The angle at the North Pole is $13.4000^o - 9.5167^o = 3.8833^o$.

See the illustration on the next page.

It follows that

$$\cos \text{n} = \cos(42.8667^o)\cos(37.4833^o) + \sin(42.8667^o)\ \sin(37.4833^o)\cos(3.8833^o)$$

$$\cos \text{n} \approx 0.994638715367 \ \Rightarrow \ \text{n} \approx \cos^{-1}(0.994638715367) \ \Rightarrow \ \text{n} \approx 5.93562267135^o.$$

Therefore, the approximate distance between the two capital cities is

$$(5.93562267135^o)\left({}^{\pi}/_{180^o}\right)(3959 \text{ miles})$$

$$\approx (0.103596158771)(3959 \text{ miles})$$

$$\approx \ 410 \ \text{ miles}.$$

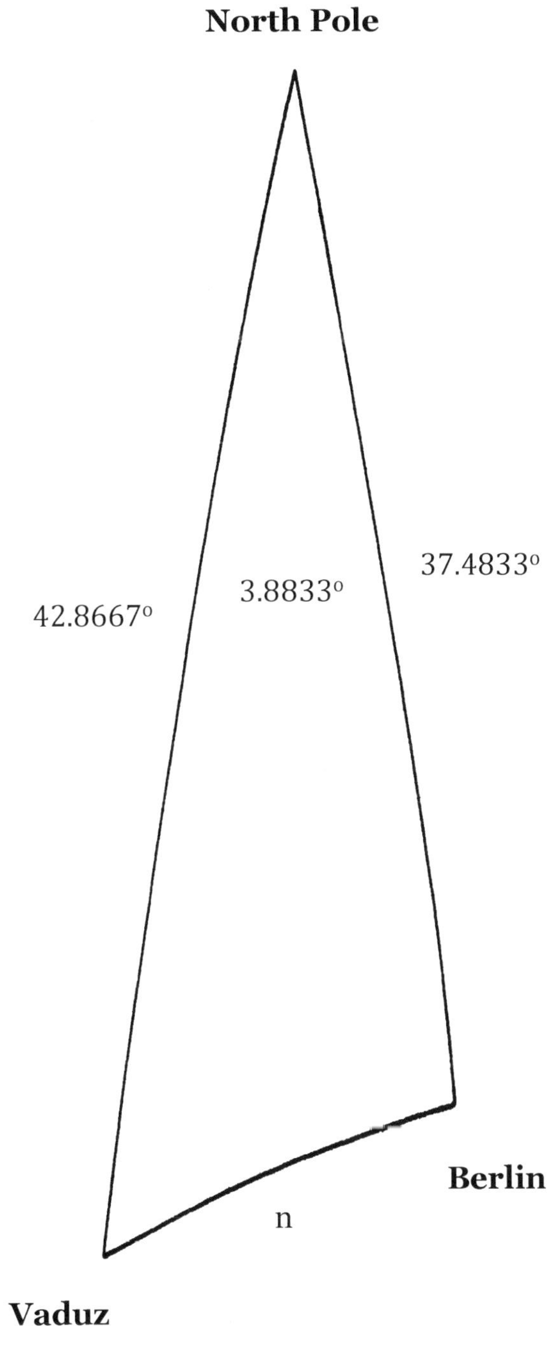
North Pole
37.4833°
3.8833°
42.8667°
Berlin
n
Vaduz

Chapter 6

Problem 16

Solution

Luxembourg, Luxembourg is located at $49^o36'$ N $= \left(49 + {}^{36}/_{60}\right)^o$ N $= 49.6000^o$ N ,

$6^o7'$ E $= \left(6 + {}^{7}/_{60}\right)^o$ E $\approx 6.1167^o$ E .

Chisinau, Moldova is located at $47^o0'$ N $= 47^o$ N ,

$28^o51'$ E $= \left(28 + {}^{51}/_{60}\right)^o$ E $= 28.8500^o$ E .

Two sides of the spherical triangle are $90^o - 49.6000^o = 40.4000^o$ and $90^o - 47^o = 43^o$, respectively. The angle at the North Pole is $28.8500^o - 6.1167^o = 22.7333^o$.

See the illustration on the next page.

It follows that

$\cos n = \cos(40.4000^o)\cos(43^o) + \sin(40.4000^o)\ \sin(43^o)\cos(22.7333^o)$

$\cos n \approx 0.964631903712 \Rightarrow n \approx \cos^{-1}(0.964631903712) \Rightarrow n \approx 15.2838183747^o$.

Therefore, the approximate distance between the two capital cities is $(15.2838183747^o)\left({}^{\pi}/_{180^o}\right)(3959 \text{ miles})$

$\approx (0.266752952915)(3959 \text{ miles})$

$\approx$ 1,056 miles.

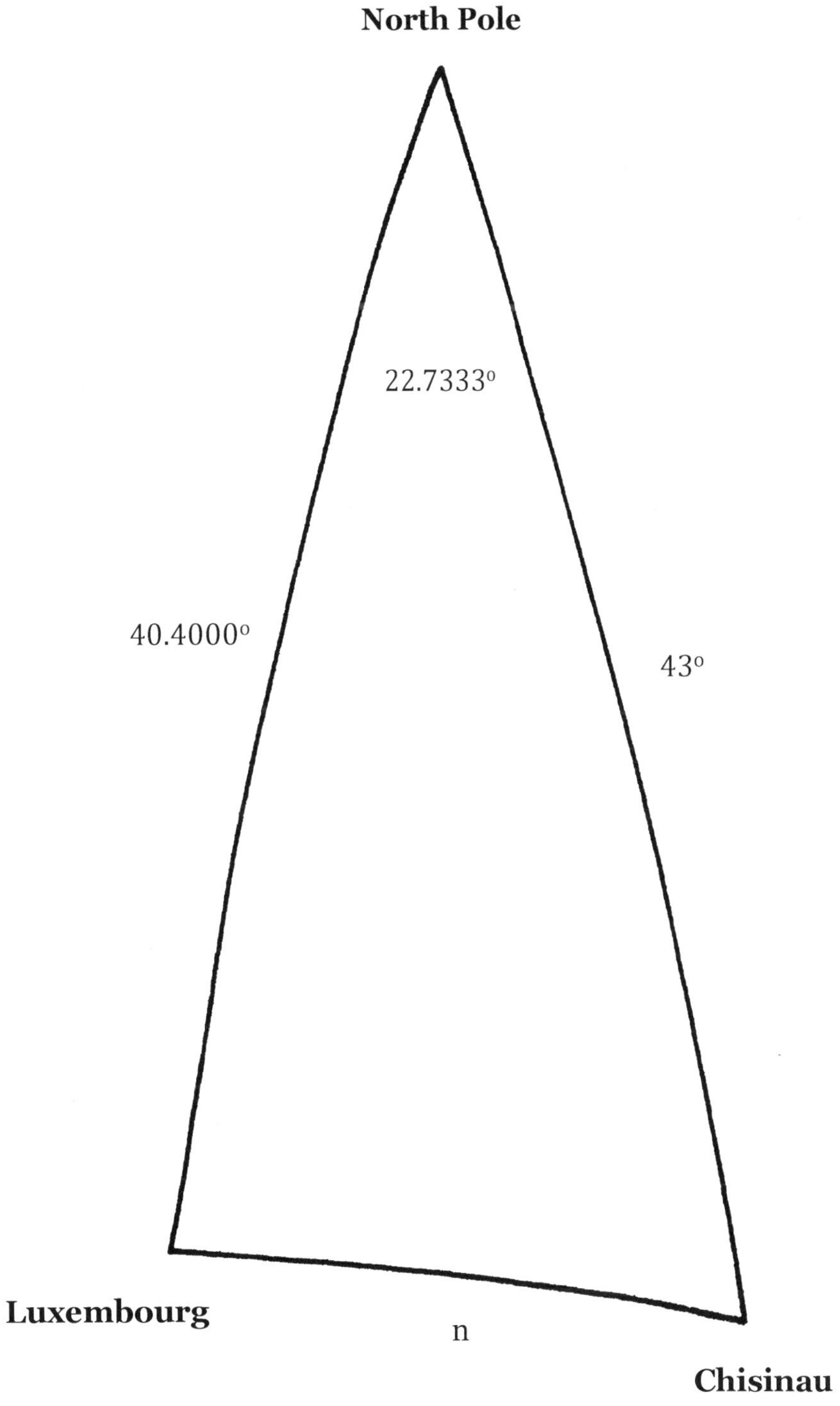
North Pole
22.7333°
40.4000°
43°
Luxembourg
n
Chisinau

Chapter 6
Problem 17
Solution

Amsterdam, Netherlands is located at $52^o21' \text{ N} = \left(52 + {}^{21}/_{60}\right)^o \text{ N} = 52.3500^o \text{ N}$,
$4^o55' \text{ E} = \left(4 + {}^{55}/_{60}\right)^o \text{ E} \approx 4.9167^o \text{ E}$.

Ljubljana, Slovenia is located at $46^o3' \text{ N} = \left(46 + {}^{3}/_{60}\right)^o \text{ N} = 46.0500^o \text{ N}$,
$14^o31' \text{ E} = \left(14 + {}^{31}/_{60}\right)^o \text{ E} \approx 14.5167^o \text{ E}$.

Two sides of the spherical triangle are $90^o - 52.3500^o = 37.6500^o$ and $90^o - 46.0500^o = 43.9500^o$, respectively. The angle at the North Pole is $14.5167^o - 4.9167^o = 9.6000^o$.

See the illustration on the next page.

It follows that

$$\cos n = \cos(37.6500^o)\cos(43.9500^o) + \sin(37.6500^o)\sin(43.9500^o)\cos(9.6000^o)$$

$$\cos n \approx 0.988024129927 \Rightarrow n \approx \cos^{-1}(0.988024129927) \Rightarrow n \approx 8.87616862272^o.$$

Therefore, the approximate distance between the two capital cities is

$$(8.87616862272^o)\left({}^{\pi}/_{180^o}\right)(3959 \text{ miles})$$

$$\approx (0.154918367429)(3959 \text{ miles})$$

$$\approx 613 \text{ miles}.$$

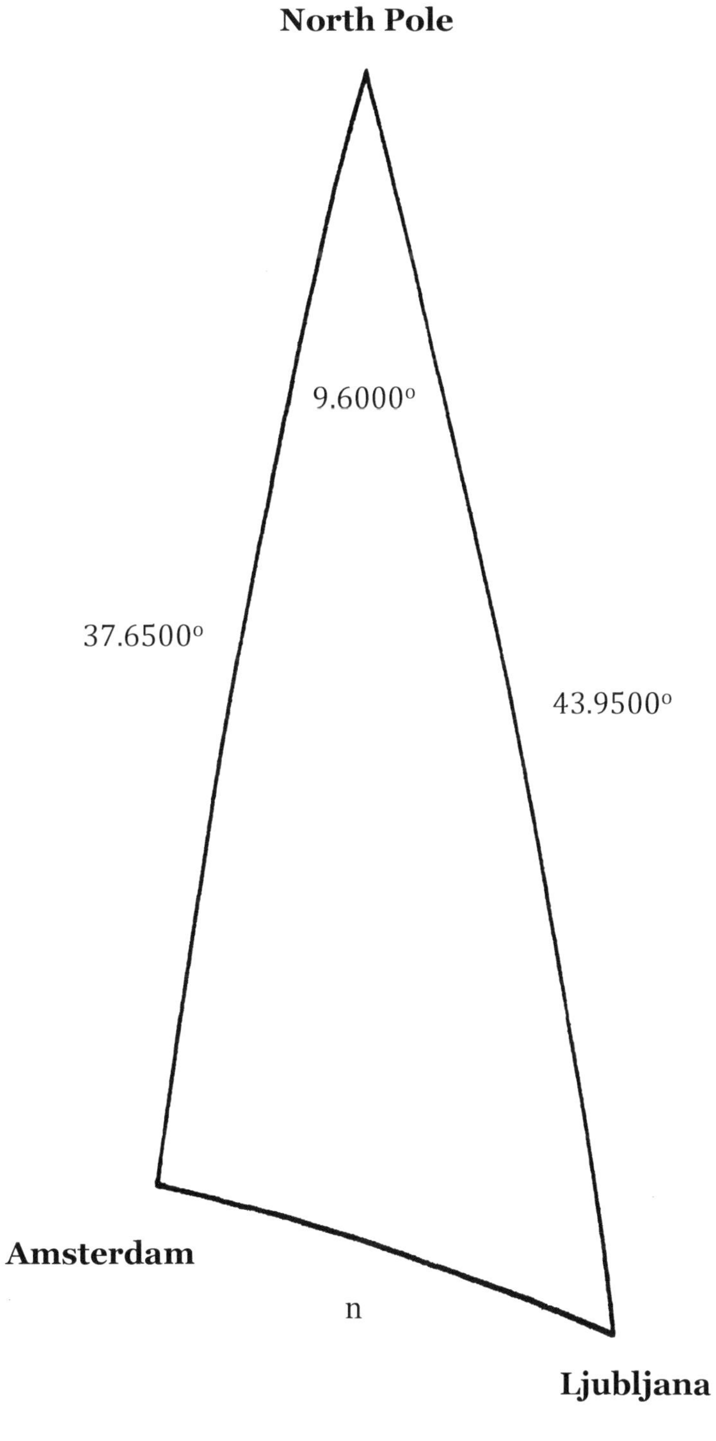
North Pole
9.6000°
37.6500°
43.9500°
Amsterdam
n
Ljubljana

Chapter 6

Problem 18

Solution

Oslo, Norway is located at $59^o55'$ N $= \left(59 + {}^{55}/_{60}\right)^o$ N $\approx 59.9167^o$ N ,
$10^o45'$ E $= \left(10 + {}^{45}/_{60}\right)^o$ E $= 10.7500^o$ E .

Athens, Greece is located at $37^o59'$ N $= \left(37 + {}^{59}/_{60}\right)^o$ N $\approx 37.9833^o$ N ,
$23^o44'$ E $= \left(23 + {}^{44}/_{60}\right)^o$ E $\approx 23.7333^o$ E .

Two sides of the spherical triangle are $90^o - 59.9167^o = 30.0833^o$ and $90^o - 37.9833^o = 52.0167^o$, respectively. The angle at the North Pole is $23.7333^o - 10.7500^o = 12.9833^o$.

See the illustration on the next page.

It follows that

$$\cos n = \cos(30.0833^o)\cos(52.0167^o) + \sin(30.0833^o)\ \sin(52.0167^o)\cos(12.9833^o)$$

$$\cos n \approx\ 0.91751849913 \Rightarrow\ n \approx \cos^{-1}(0.91751849913)\ \Rightarrow\ n \approx 23.4340421661^o.$$

Therefore, the approximate distance between the two capital cities is

$(23.4340421661^o)\left({}^{\pi}/_{180^o}\right)(3959 \text{ miles})$

$\approx (0.40900119285)(3959 \text{ miles})$

1,619 miles.

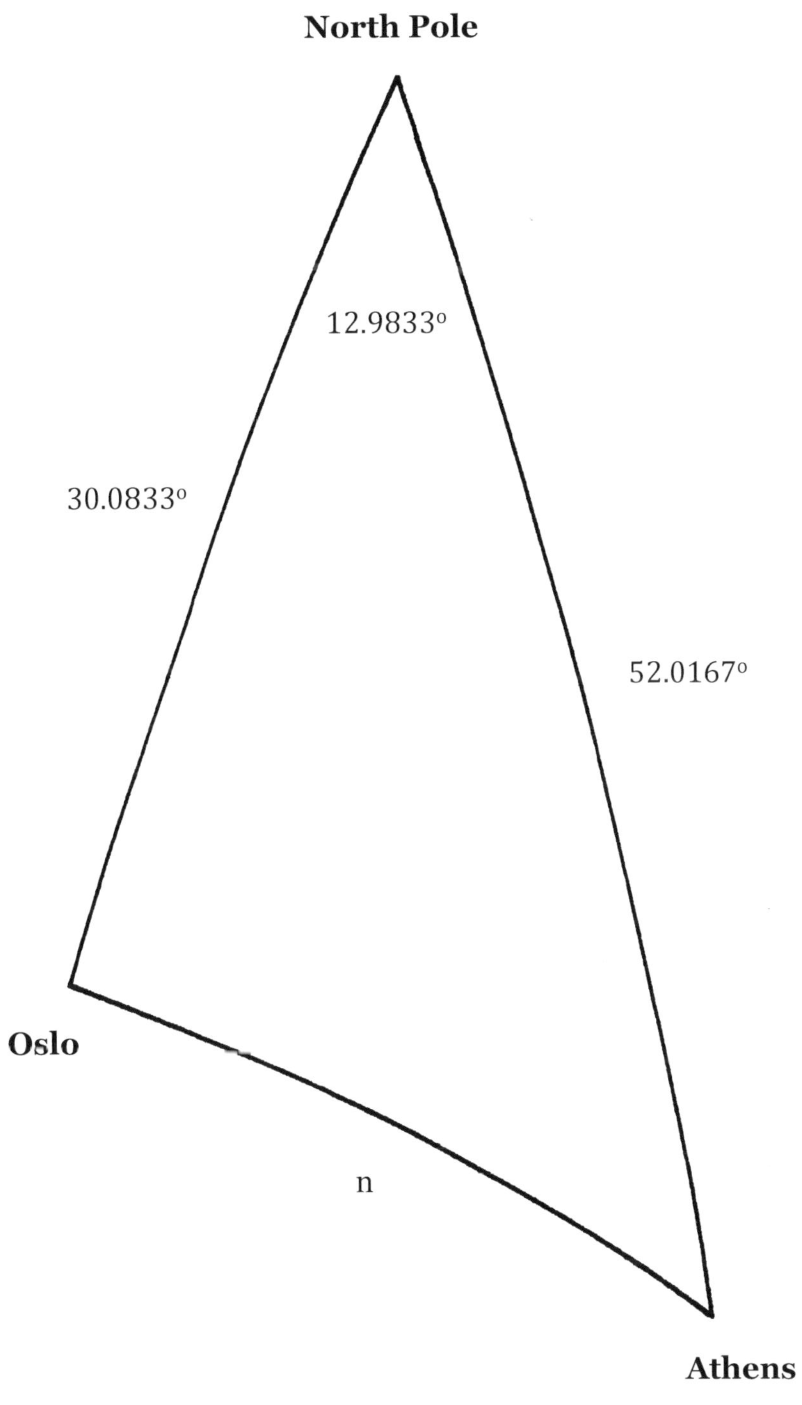
North Pole
12.9833°
30.0833°
52.0167°
Oslo
n
Athens

Chapter 6

Problem 19

Solution

Warsaw, Poland is located at $52^o 15' \text{ N} = \left(52 + {}^{15}/_{60}\right)^o \text{ N} = 52.2500^o \text{ N}$, $21^o 0' \text{ E} = 21^o \text{ E}$

Nicosia, Cyprus is located at $35^o 10' \text{ N} = \left(35 + {}^{10}/_{60}\right)^o \text{ N} \approx 35.1667^o \text{ N}$, $33^o 22' \text{ E} = \left(33 + {}^{22}/_{60}\right)^o \text{ E} \approx 33.3667^o \text{ E}$.

The sides of the spherical triangle are $90^o - 52.2500^o = 37.7500^o$ and $90^o - 35.1667^o = 54.8333^o$, respectively. The angle at the North Pole is $33.3667^o - 21^o = 12.3667^o$.

See the illustration on the next page.

It follows that

$\cos n = \cos(37.7500^o)\cos(54.8333^o) + \sin(37.7500^o)\ \sin(54.8333^o)\cos(12.3667^o)$

$\cos n \approx 0.944266109124 \ \Rightarrow\ n \approx \cos^{-1}(0.944266109124) \ \Rightarrow\ n \approx 19.219200665^o$.

Therefore, the approximate distance between the two capital cities is

$(19.219200665^o)\left({}^{\pi}/_{180^o}\right)(3959 \text{ miles})$

$\approx (0.335438331206)(3959 \text{ miles})$

$\approx\ 1{,}328\ \text{ miles}$.

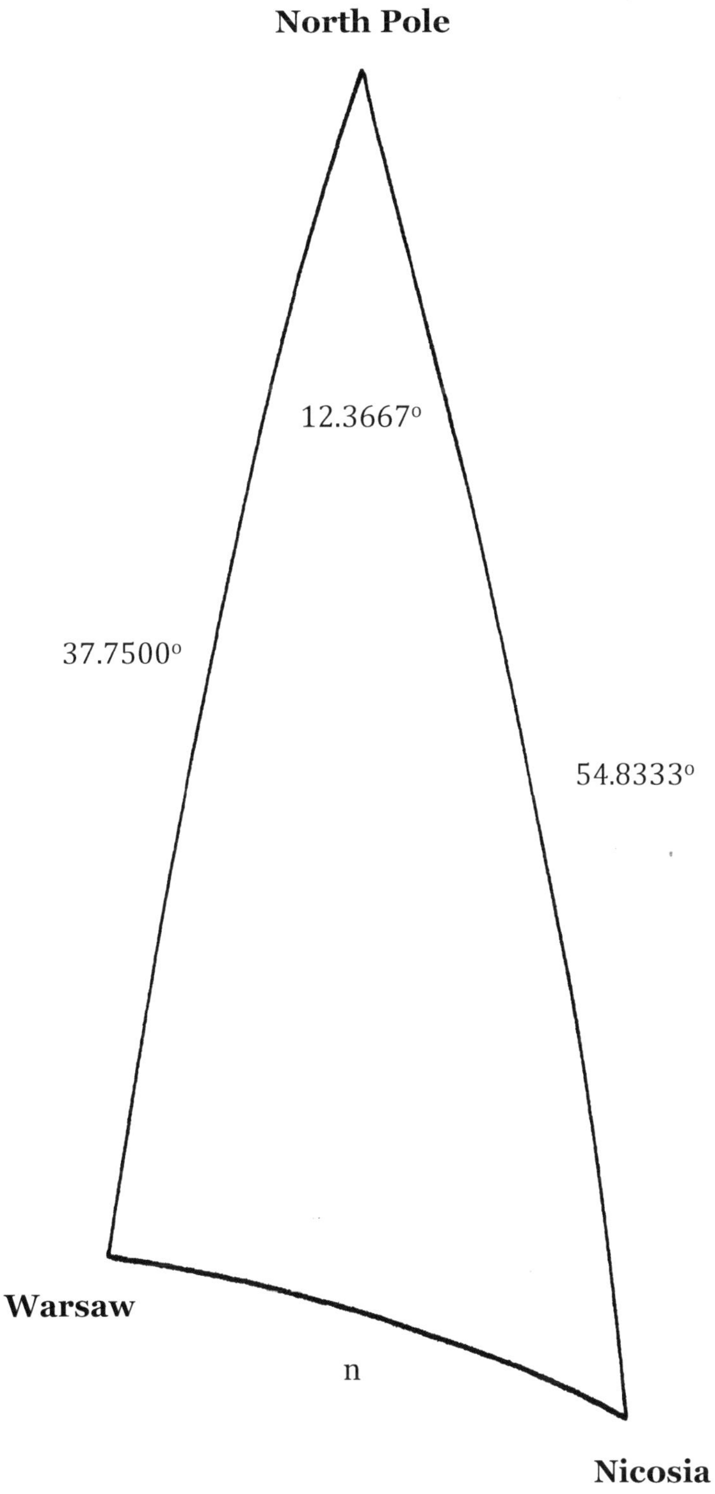
North Pole
12.3667°
37.7500°
54.8333°
Warsaw
n
Nicosia

Chapter 6
Problem 20
Solution

San Marino, San Marino is located at $43^o56'\text{ N} = \left(43 + {}^{56}/_{60}\right)^o \text{ N} \approx 43.9333^o \text{ N}$,
$12^o25'\text{ E} = \left(12 + {}^{25}/_{60}\right)^o \text{ E} \approx 12.4167^o \text{ E}$.

Kyiv, Ukraine is located at $50^o26'\text{ N} = \left(50 + {}^{26}/_{60}\right)^o \text{ N} \approx 50.4333^o \text{ N}$,
$30^o31'\text{ E} = \left(30 + {}^{31}/_{60}\right)^o \text{ E} \approx 30.5167^o \text{ E}$.

Two sides of the spherical triangle are $90^o - 43.9333^o = 46.0667^o$ and $90^o - 50.4333^o = 39.5667^o$, respectively. The angle at the North Pole is $30.5167^o - 12.4167^o = 18.1000^o$.

See the illustration on the next page.

It follows that

$\cos n = \cos(46.0667^o)\cos(39.5667^o) + \sin(46.0667^o)\ \sin(39.5667^o)\cos(18.1000^o)$

$\cos n \approx 0.970872578996 \ \Rightarrow\ n \approx \cos^{-1}(0.970872578996) \ \Rightarrow\ n \approx 13.8627208532^o.$

Therefore, the approximate distance between the two capital cities is
$(13.8627208532^o)\left({}^{\pi}/_{180^o}\right)(3959 \text{ miles})$

$\approx (0.241950122172)(3959 \text{ miles})$

$\approx\ 958\ \text{ miles}.$

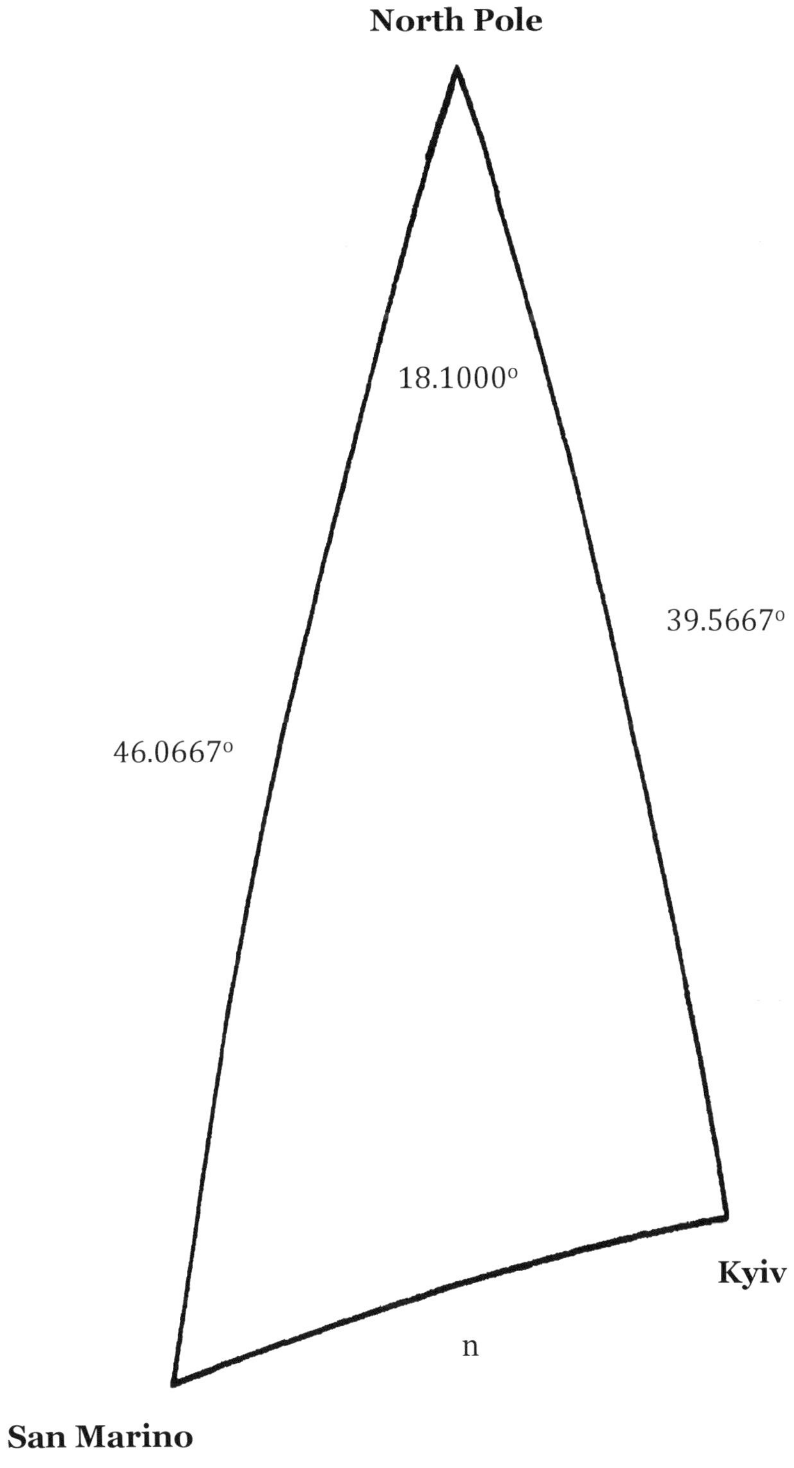
North Pole
18.1000°
39.5667°
46.0667°
Kyiv
n
San Marino

Chapter 6

Problem 21

Solution

Belgrade, Serbia is located at $44^o50'\text{ N} = \left(44 + {}^{50}/_{60}\right)^o \text{ N} \approx 44.8333^o \text{ N}$, $20^o30'\text{ E} = \left(20 + {}^{30}/_{60}\right)^o \text{ E} = 20.5000^o \text{ E}$

Helsinki, Finland is located at $60^o10'\text{ N} = \left(60 + {}^{10}/_{60}\right)^o \text{ N} \approx 60.1667^o \text{ N}$, $24^o56'\text{ E} = \left(24 + {}^{56}/_{60}\right)^o \text{ E} \approx 24.9333^o \text{ E}$

The sides of the spherical triangle are $90^o - 44.8333^o = 45.1667^o$ and $90^o - 60.1667^o = 29.8333^o$, respectively. The angle at the North Pole is $24.9333^o - 20.5000^o = 4.4333^o$.

See the illustration on the next page.

It follows that

$\cos n = \cos(45.1667^o)\cos(29.8333^o) + \sin(45.1667^o)\ \sin(29.8333^o)\cos(4.4333^o)$

$\cos n \approx 0.963347876475 \ \Rightarrow\ n \approx \cos^{-1}(0.963347876475) \ \Rightarrow\ n \approx 15.5604692661^o.$

Therefore, the approximate distance between the two capital cities is

$(15.5604692661^o)\left({}^{\pi}/_{180^o}\right)(3959 \text{ miles})$

$\approx (0.271581421849)(3959 \text{ miles})$

$\approx\ 1{,}075\ \text{ miles}.$

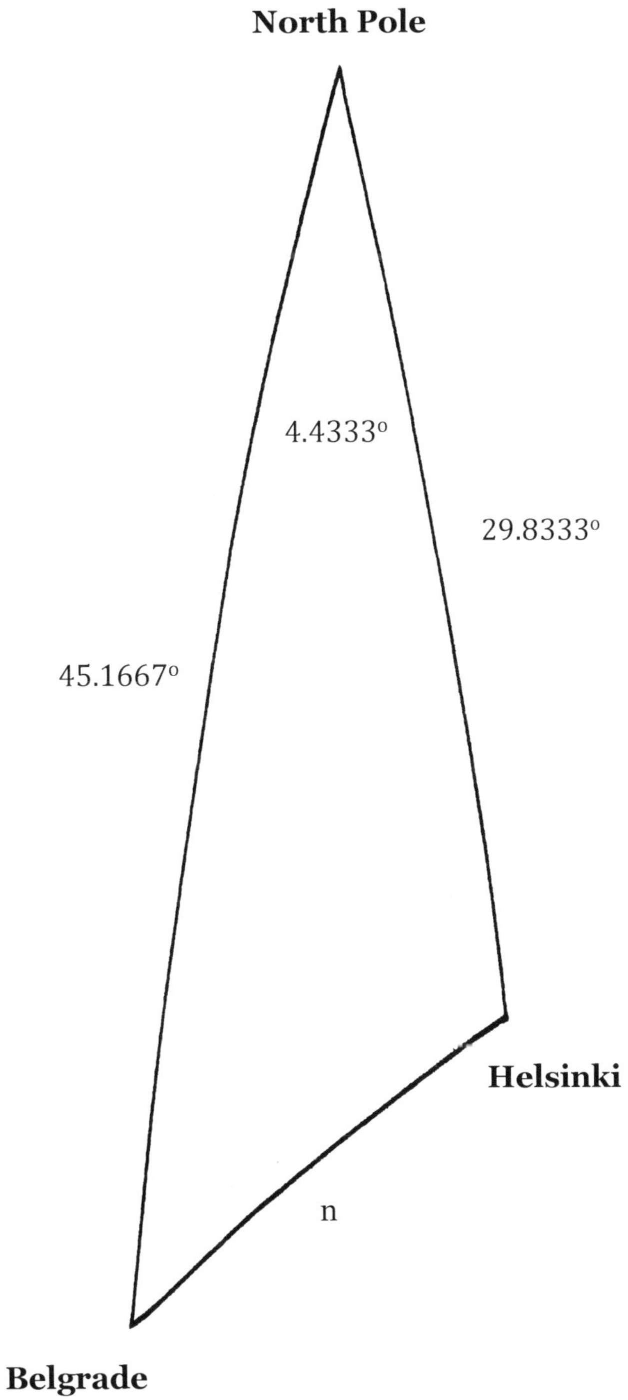
North Pole
4.4333°
29.8333°
45.1667°
Helsinki
n
Belgrade

Chapter 6
Problem 22
Solution

Bern, Switzerland is located at $46^o 55' \text{ N} = \left(46 + {}^{55}/_{60}\right)^o \text{ N} \approx 46.9167^o \text{ N}$,

$7^o 28' \text{ E} = \left(7 + {}^{28}/_{60}\right)^o \text{ E} \approx 7.4667^o \text{ E}$.

Valletta, Malta is located at $35^o 53' \text{ N} = \left(35 + {}^{53}/_{60}\right)^o \text{ N} \approx 35.8833^o \text{ N}$,

$14^o 30' \text{ E} = \left(14 + {}^{30}/_{60}\right)^o \text{ E} = 14.5000^o \text{ E}$.

Two sides of the spherical triangle are $90^o - 46.9167^o = 43.0833^o$ and $90^o - 35.8833^o = 54.1167^o$, respectively. The angle at the North Pole is $14.5000^o - 7.4667^o = 7.0333^o$.

See the illustration on the next page.

It follows that

$\cos n = \cos(43.0833^o)\cos(54.1167^o) + \sin(43.0833^o)\ \sin(54.1167^o)\cos(7.0333^o)$

$\cos n \approx 0.977351352377 \;\Rightarrow\; n \approx \cos^{-1}(0.977351352377) \;\Rightarrow\; n \approx 12.2174866883^o$.

Therefore, the approximate distance between the two capital cities is

$(12.2174866883^o)\left({}^{\pi}/_{180^o}\right)(3959 \text{ miles})$

$\approx (0.213235369029)(3959 \text{ miles})$

$\approx\ 844\ \text{ miles}$.

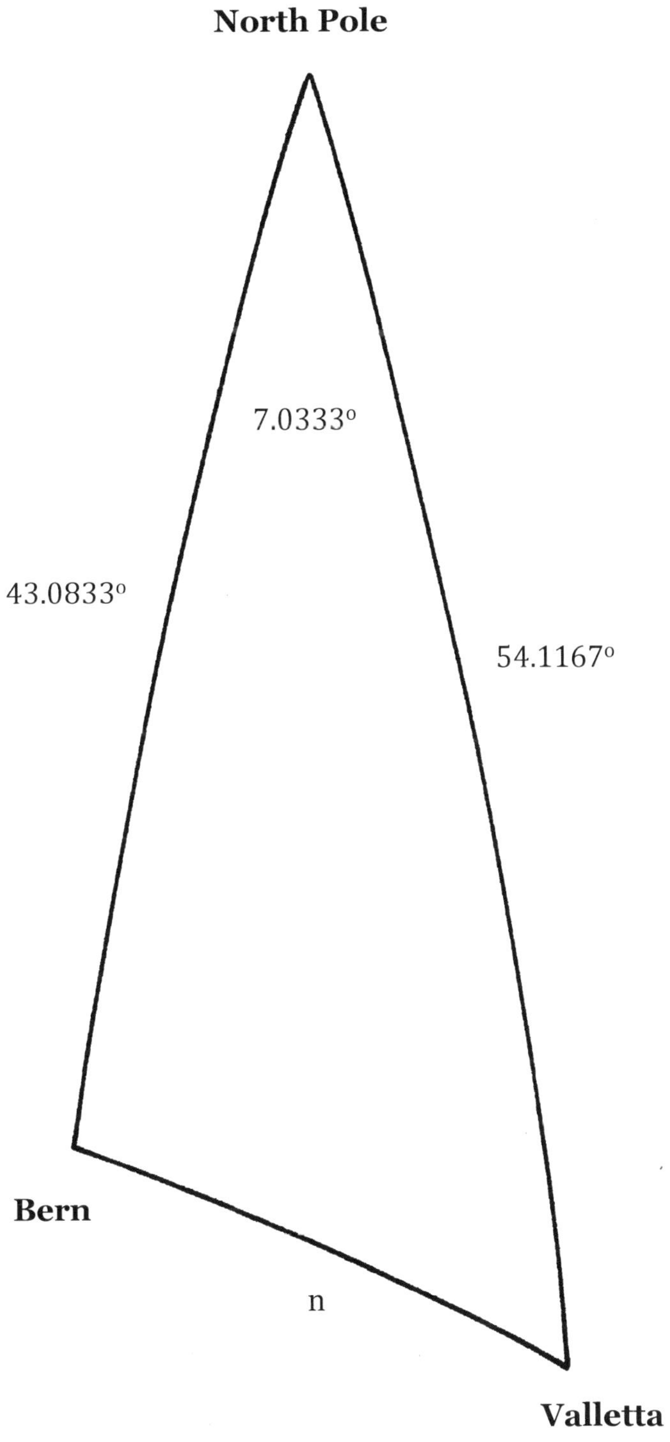
North Pole
7.0333°
43.0833°
54.1167°
Bern
n
Valletta

Chapter 6

Problem 23

Solution

Longyearbyen, Svalbard is located at $78^o 13' \text{ N} = \left(78 + {}^{13}/_{60}\right)^o \text{ N} \approx 78.2167^o \text{ N}$, $15^o 38' \text{ E} = \left(15 + {}^{38}/_{60}\right)^o \text{ E} \approx 15.6333^o \text{ E}$.

Pristina, Kosovo is located at $42^o 40' \text{ N} = \left(42 + {}^{40}/_{60}\right)^o \text{ N} \approx 42.6667^o \text{ N}$, $21^o 10' \text{ E} = \left(21 + {}^{10}/_{60}\right)^o \text{ E} \approx 21.1667^o \text{ E}$.

Two sides of the spherical triangle are $90^o - 78.2167^o = 11.7833^o$ and $90^o - 42.6667^o = 47.3333^o$, respectively. The angle at the North Pole is $21.1667^o - 15.6333^o = 5.5334^o$.

See the illustration on the next page.

It follows that

$$\cos n = \cos(11.7833^o)\cos(47.3333^o) + \sin(11.7833^o)\ \sin(47.3333^o)\cos(5.5334^o)$$

$$\cos n \approx 0.8129087372 \ \Rightarrow \ n \approx \cos^{-1}(0.8129087372) \ \Rightarrow \ n \approx 35.618895703^o.$$

Therefore, the approximate distance between the two capital cities is

$$(35.618895703^o)\left({}^{\pi}/_{180^o}\right)(3959 \text{ miles})$$

$$\approx (0.621667005941)(3959 \text{ miles})$$

$$\approx \ 2{,}461 \ \text{ miles.}$$

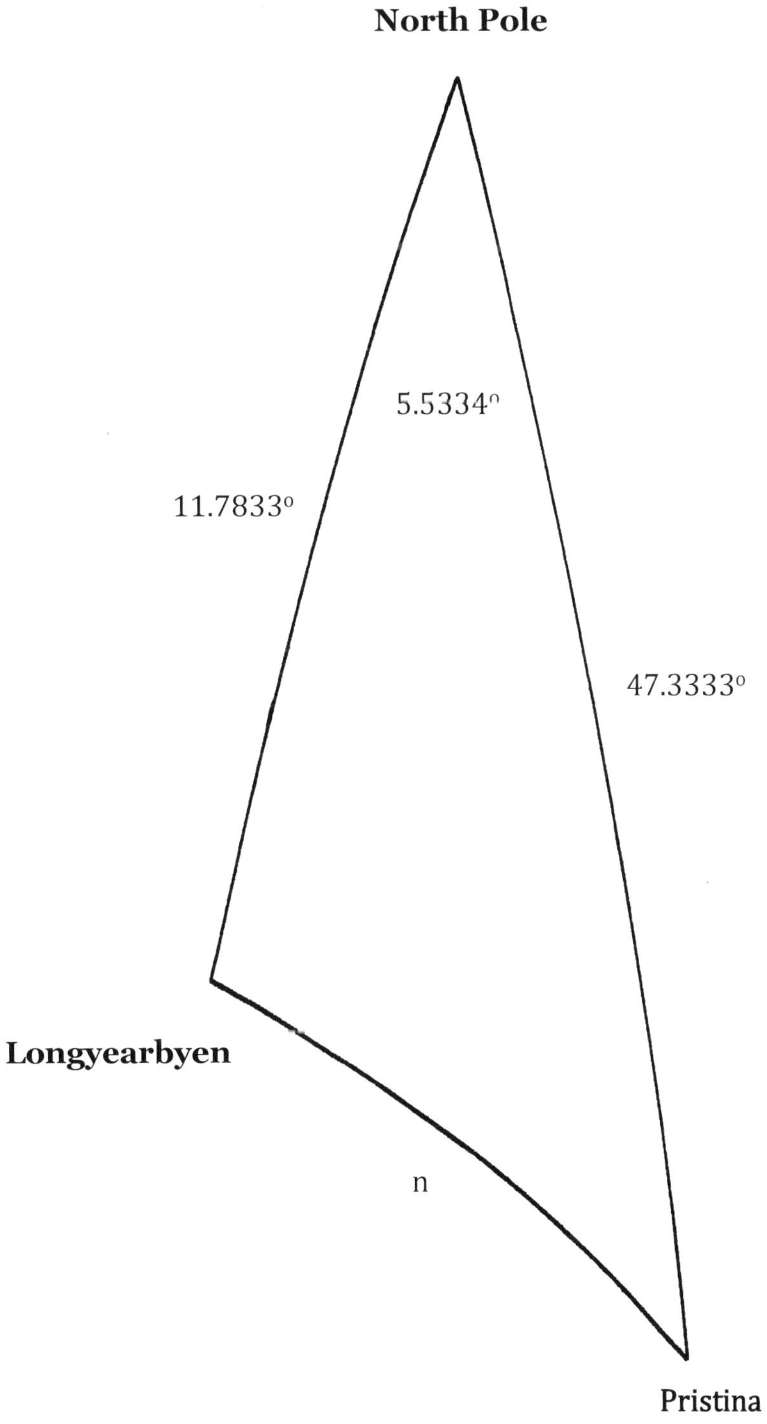
North Pole
5.5334°
11.7833°
47.3333°
Longyearbyen
n
Pristina

Chapter 6
Problem 24
Solution

London, United Kingdom is located at $51^o30' \text{ N} = \left(51 + {}^{30}/_{60}\right)^o \text{ N} = 51.5000^o \text{ N}$,

$0^o5' \text{ W} = \left({}^{5}/_{60}\right)^o \text{ W} \approx 0.0833^o \text{ W}$.

Bratislava, Slovakia is located at $48^o9' \text{ N} = \left(48 + {}^{9}/_{60}\right)^o \text{ N} = 48.1500^o \text{ N}$,

$17^o7' \text{ E} = \left(17 + {}^{7}/_{60}\right)^o \text{ E} \approx 17.1167^o \text{ E}$.

Two sides of the spherical triangle are $90^o - 51.5000^o = 38.5000^o$ and $90^o - 48.1500^o = 41.8500^o$, respectively. The angle at the North Pole is $17.1167^o + 0.0833^o = 17.2000^o$.

See the illustration on the next page.

It follows that

$\cos n = \cos(38.5000^o)\cos(41.8500^o) + \sin(38.5000^o)\,\sin(41.8500^o)\cos(17.2000^o)$

$\cos n \approx 0.9797169181 \Rightarrow n \approx \cos^{-1}(0.9797169181) \Rightarrow n \approx 11.559563^o$.

Therefore, the approximate distance between the two capital cities is

$(11.559563^o)\left({}^{\pi}/_{180^o}\right)(3959 \text{ miles})$

$\approx (0.2017524345)(3959 \text{ miles})$

≈ 799 miles.

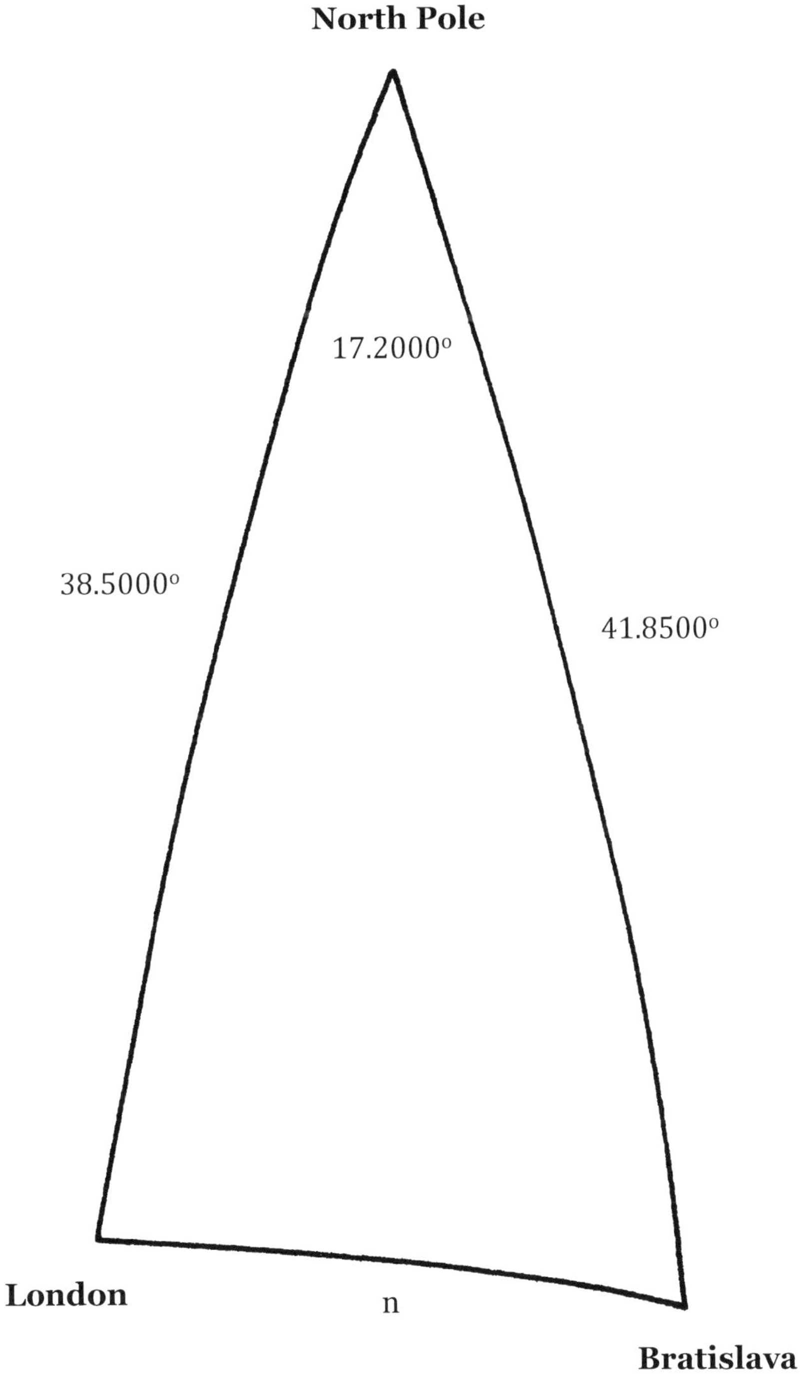
North Pole
17.2000°
38.5000°
41.8500°
London
n
Bratislava

Chapter 7

Middle East

Example 1

Amman, Jordan is located at

$31^o57'\ \text{N} = \left(31 + {}^{57}/_{60}\right)^o\ \text{N} = 31.9500^o\ \text{N}$,

$35^o56'\ \text{E} = \left(35 + {}^{56}/_{60}\right)^o\ \text{E} \approx 35.9333^o\ \text{E}$.

Riyadh, Saudi Arabia is located at

$24^o39'\ \text{N} = \left(24 + {}^{39}/_{60}\right)^o\ \text{N} = 24.6500^o\ \text{N}$,

$46^o42'\ \text{E} = \left(46 + {}^{42}/_{60}\right)^o\ \text{E} = 46.7000^o\ \text{E}$.

Two sides of the spherical triangle are $90^o - 31.9500^o = 58.0500^o$ and $90^o - 24.6500^o = 65.3500^o$, respectively. The angle at the North Pole is $46.7000^o - 35.9333^o = 10.7667^o$.

See the illustration on the next page.

It follows that

$\cos \text{n} = \cos(58.0500^o)\cos(65.3500^o) + \sin(58.0500^o)\ \sin(65.3500^o)\cos(10.7667^o)$

$\cos \text{n} \approx 0.978318448394\ \Rightarrow\ \text{n} \approx \cos^{-1}(0.978318448394)\ \Rightarrow\ \text{n} \approx 11.9528268659^o$.

Therefore, the approximate distance between the two capital cities is

$(11.9528268659^o)\left({}^{\pi}/_{180^o}\right)(3959\ \text{miles})$

$\approx (0.208616183731)(3959\ \text{miles})$

≈ 826 miles.

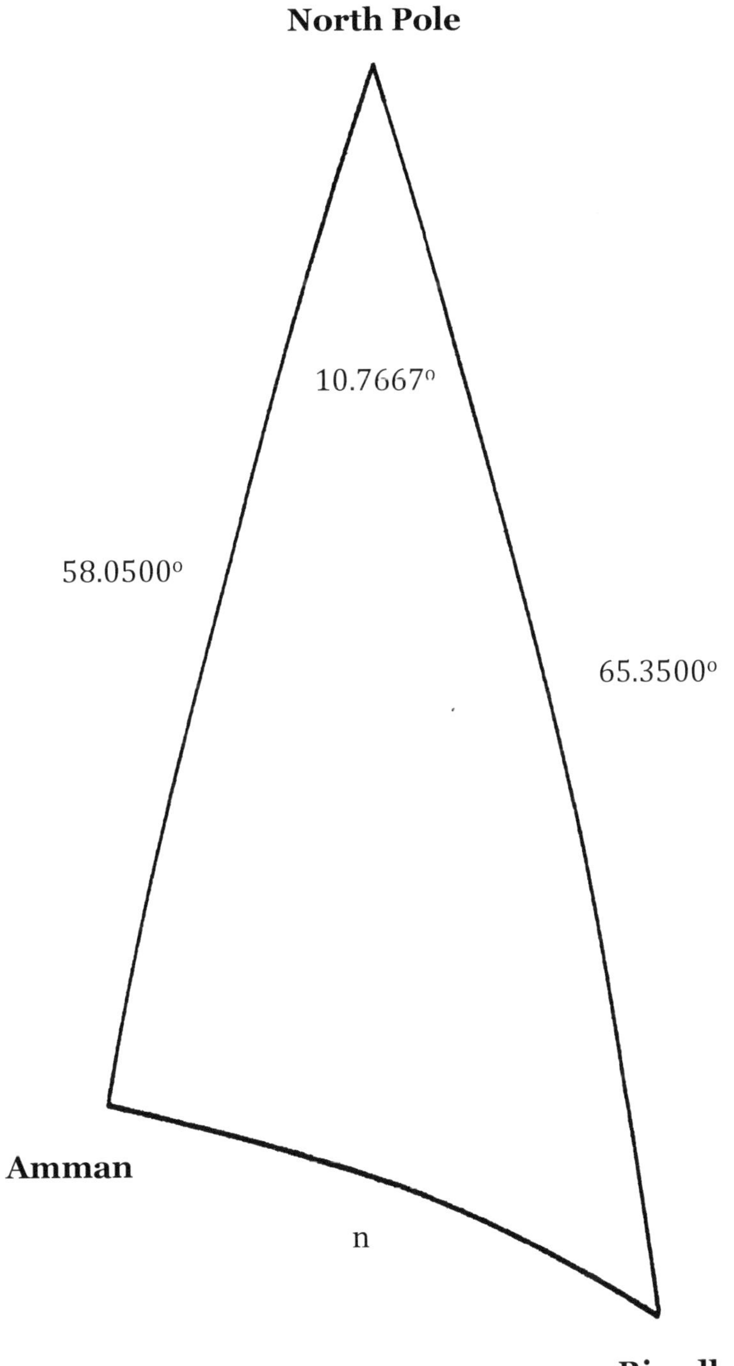
North Pole
10.7667°
58.0500°
65.3500°
Amman
n
Riyadh

Chapter 7 Problems:

Approximate the distance between the following capitals in the Middle East.

1. Yerevan, Armenia (40°10′ N 44°30′ E)
   to Manama, Bahrain (26°14′ N 50°34′ E)

2. Tbilisi, Georgia (41°41′ N 44°50′ E)
   to Baku, Azerbaijan (40°23′ N 49°52′ E)

3. Jerusalem, Israel (31°46′ N 35°14′ E) to
   Tbilisi, Georgia (41°41′ N 44°50′ E)

4. Beirut, Lebanon (33°52′ N 35°30′ E)
   to Kuwait City, Kuwait (29°22′ N 47°58′ E)

5. Damascus, Syria (33°30′ N 36°18′ E)
   to Doha, Qatar (25°17′ N 51°32′ E)

6. Abu Dhabi, United Arab Emirates (24°28′ N 54°22′ E)
   to Muscat, Oman (23°37′ N 58°35′ E)

7. Ankara, Turkey (39°56′ N 32°52′ E)
   to Baghdad, Iraq (33°20′ N 44°24′ E)

8. Sanaa, Yemen (15°21′ N 44°12′ E)
   to Tehran, Iran (35°42′ N 51°25′ E)

Chapter 7

Problem 1

Solution

Yerevan, Armenia is located at

$40^o10'\ \text{N} = \left(40 + {}^{10}/_{60}\right)^o\ \text{N} \approx 40.1667^o\ \text{N}$,

$44^o30'\ \text{E} = \left(44 + {}^{30}/_{60}\right)^o\ \text{E} = 44.5000^o\ \text{E}$.

Manama, Bahrain is located at

$26^o14'\ \text{N} = \left(26 + {}^{14}/_{60}\right)^o\ \text{N} \approx 26.2333^o\ \text{N}$,

$50^o34'\ \text{E} = \left(50 + {}^{34}/_{60}\right)^o\ \text{E} \approx 50.5667^o\ \text{E}$.

Two sides of the spherical triangle are $90^o - 40.1667^o = 49.8333^o$ and $90^o - 26.2333^o = 63.7667^o$, respectively. The angle at the North Pole is $50.5667^o - 44.5000^o = 6.0667^o$.

See the illustration on the next page.

It follows that

$\cos n = \cos(49.8333^o)\cos(63.7667^o) + \sin(49.8333^o)\ \sin(63.7667^o)\cos(6.0667^o)$

$\cos n \approx 0.966737369406 \quad \Rightarrow n \approx \cos^{-1}(0.966737369406\,) \Rightarrow n \approx 14.81928159^o$.

Therefore, the approximate distance between the two capital cities is

$(14.81928159^o)\left({}^{\pi}/_{180^o}\right)(3959\ \text{miles})$

$\approx (0.258645256525)(3959\ \text{miles})$

$\approx\ 1{,}024\ \text{miles}$.

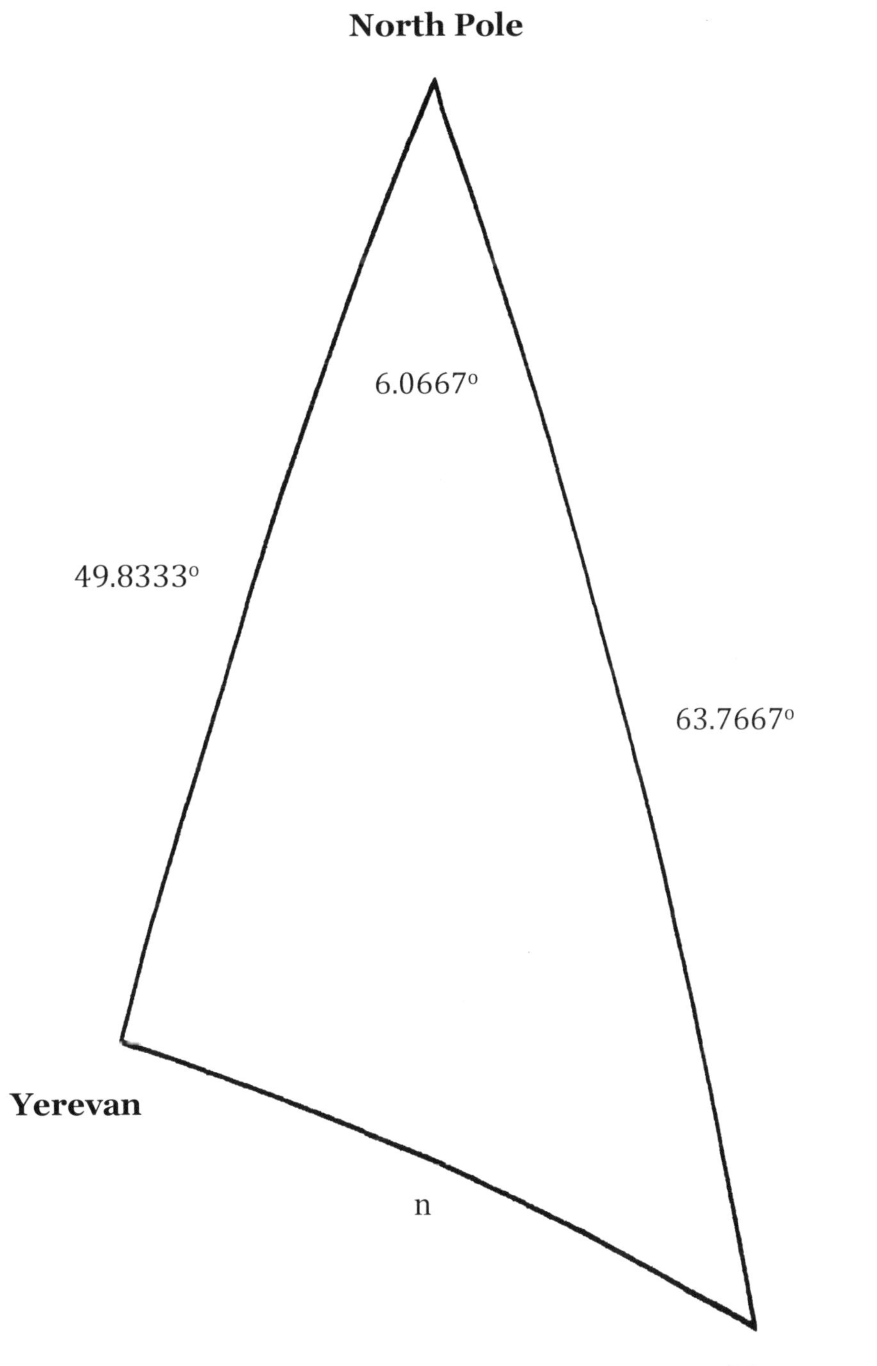
North Pole
6.0667°
49.8333°
63.7667°
Yerevan
n
Manama

Chapter 7

Problem 2

Solution

Tbilisi, Georgia is located at $41^o41'\ \mathrm{N} = \left(41 + {}^{41}/_{60}\right)^o\ \mathrm{N} \approx 41.6833^o\ \mathrm{N}$,

$44^o50'\ \mathrm{E} = \left(44 + {}^{50}/_{60}\right)^o\ \mathrm{E} \approx 44.8333^o\ \mathrm{E}$.

Baku, Azerbaijan is located at $40^o23'\ \mathrm{N} = \left(40 + {}^{23}/_{60}\right)^o\ \mathrm{N} \approx 40.3833^o\ \mathrm{N}$,

$49^o52'\ \mathrm{E} = \left(49 + {}^{52}/_{60}\right)^o\ \mathrm{E} \approx 49.8667^o\ \mathrm{E}$.

Two sides of the spherical triangle are $90^o - 41.6833^o = 48.3167^o$ and $90^o - 40.3833^o = 49.6167^o$, respectively. The angle at the North Pole is $49.8667^o - 44.8333^o = 5.0334^o$.

See the illustration on the next page.

It follows that

$\cos n = \cos(48.3167^o)\cos(49.6167^o) + \sin(48.3167^o)\ \sin(49.6167^o)\cos(5.0334^o)$

$\cos n \approx 0.997548841292 \ \Rightarrow\ n \approx \cos^{-1}(0.997548841292) \ \Rightarrow\ n \approx 4.01247275436^o$.

Therefore, the approximate distance between the two capital cities is

$(4.01247275436^o)\left({}^{\pi}/_{180^o}\right)(3959\ \text{miles})$

$\approx (0.07003086071)(3959\ \text{miles})$

$\approx\ 277\ \ \text{miles}.$

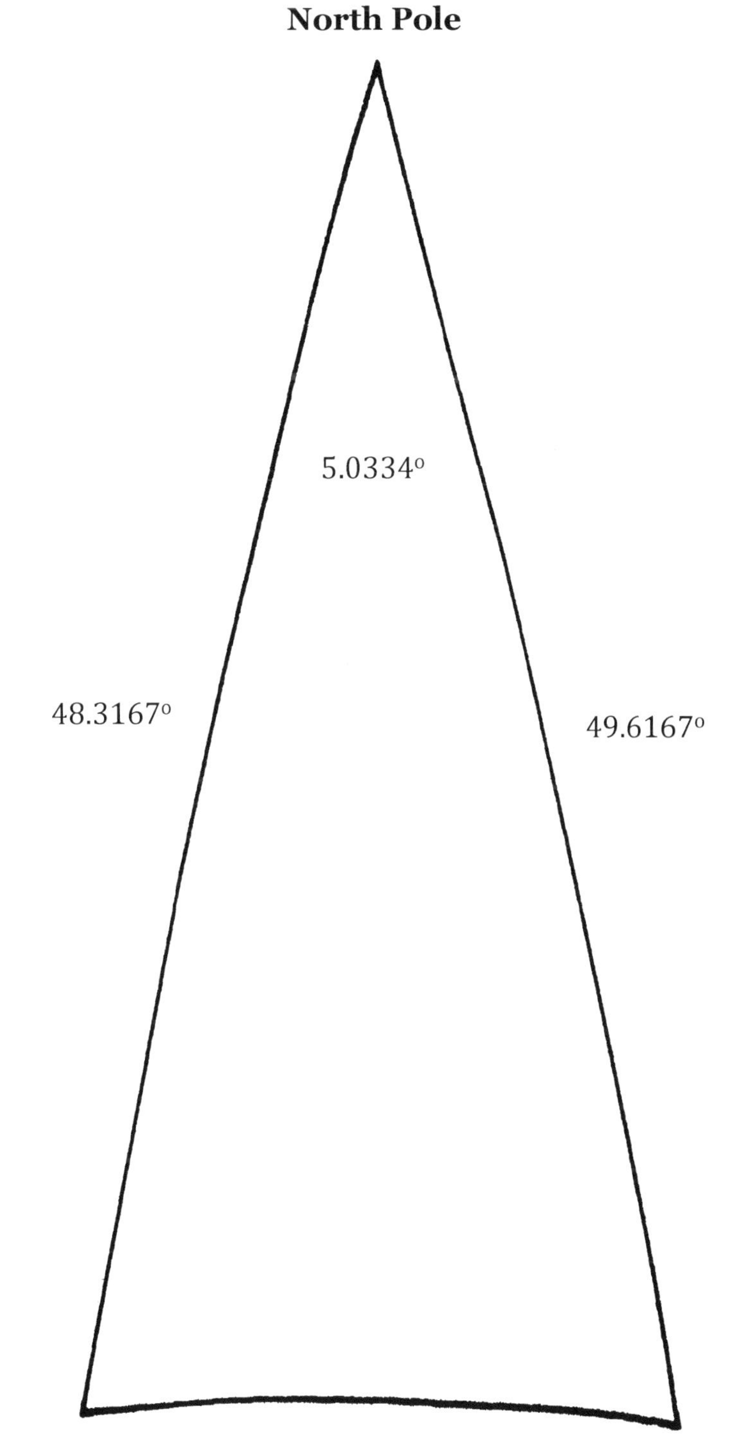
North Pole
5.0334°
48.3167°
49.6167°
Tbilisi
n
Baku

Chapter 7
Problem 3
Solution

Jerusalem, Israel is located at $31^o46'$ N $= \left(31 + {}^{46}/_{60}\right)^o$ N $\approx 31.7667^o$ N ,

$35^o14'$ E $= \left(35 + {}^{14}/_{60}\right)^o$ E $\approx 35.2333^o$ E .

Tbilisi, Georgia is located at $41^o41'$ N $= \left(41 + {}^{41}/_{60}\right)^o$ N $\approx 41.6833^o$ N ,

$44^o50'$ E $= \left(44 + {}^{50}/_{60}\right)^o$ E $\approx 44.8333^o$ E .

Two sides of the spherical triangle have measures of $90^o - 31.7667^o = 58.2333^o$ and $90^o - 41.6833^o = 48.3167^o$, respectively. The angle opposite side n has a measure of $44.8333^o - 35.2333^o = 9.6000^o$.

See the illustration on the next page.

It follows that

$\cos n = \cos(58.2333^o)\cos(48.3167^o) + \sin(58.2333^o)\ \sin(48.3167^o)\cos(9.6000^o)$

$\cos n \approx 0.9761675764 \ \Rightarrow\ n \approx \cos^{-1}(0.9761675764)\ \Rightarrow\ n \approx 12.53395275^o.$

Therefore, the approximate distance between the two capital cities is

$(12.53395275)\left({}^{\pi}/_{180^o}\right)(3959 \text{ miles})$

$\approx (0.2187587437)(3959 \text{ miles})$

$\approx\ 866\ \text{ miles}.$

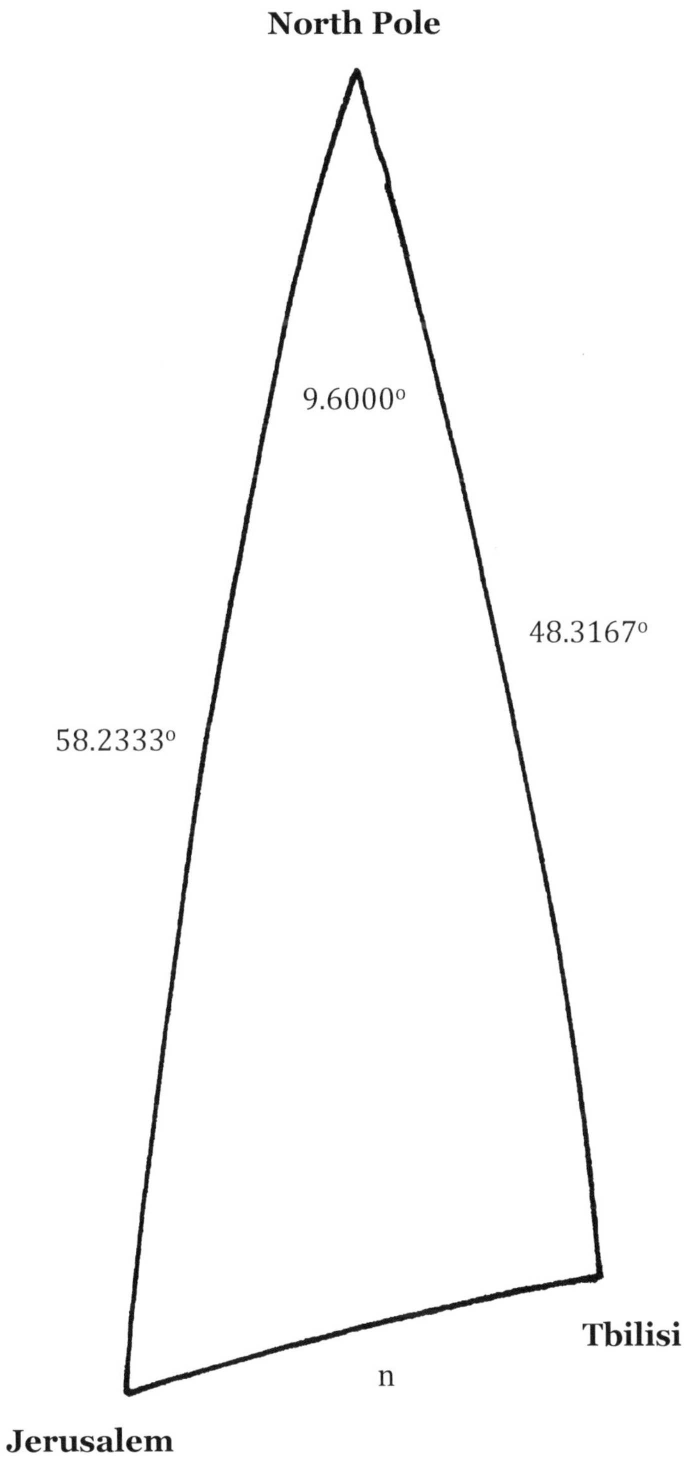
North Pole
9.6000°
48.3167°
58.2333°
Tbilisi
n
Jerusalem

Chapter 7
Problem 4
Solution

Beirut, Lebanon is located at $33^o52'\ \mathrm{N} = \left(33 + {}^{52}/_{60}\right)^o\ \mathrm{N} \approx 33.8667^o\ \mathrm{N}$,
$35^o30'\ \mathrm{E} = \left(35 + {}^{30}/_{60}\right)^o\ \mathrm{E} \approx 35.5000^o\ \mathrm{E}$.

Kuwait City, Kuwait is located at $29^o22'\ \mathrm{N} = \left(29 + {}^{22}/_{60}\right)^o\ \mathrm{N} = 29.3667^o\ \mathrm{N}$,
$47^o58'\mathrm{E} = \left(47 + {}^{58}/_{60}\right)^o\ \mathrm{E} \approx 47.9667^o\ \mathrm{E}$.

Two sides of the spherical triangle are $90^o - 33.8667^o = 56.1333^o$ and $90^o - 29.3667^o = 60.6333^o$, respectively. The angle at the North Pole is $47.9667^o - 35.5000^o = 12.4667^o$.

See the illustration on the next page.

It follows that

$\cos n = \cos(56.1333^o)\cos(60.6333^o) + \sin(56.1333^o)\ \sin(60.6333^o)\cos(12.4667^o)$

$\cos n \approx 0.9798551514 \ \Rightarrow\ n \approx \cos^{-1}(0.9798551514)\ \Rightarrow\ n \approx 11.5199716^o$.

Therefore, the approximate distance between the two capital cities is

$(11.5199716^o)\left({}^{\pi}/_{180^o}\right)(3959\text{ miles})$

$\approx (0.2010614341)(3959\text{ miles})$

$\approx\ 796\ \text{ miles}$.

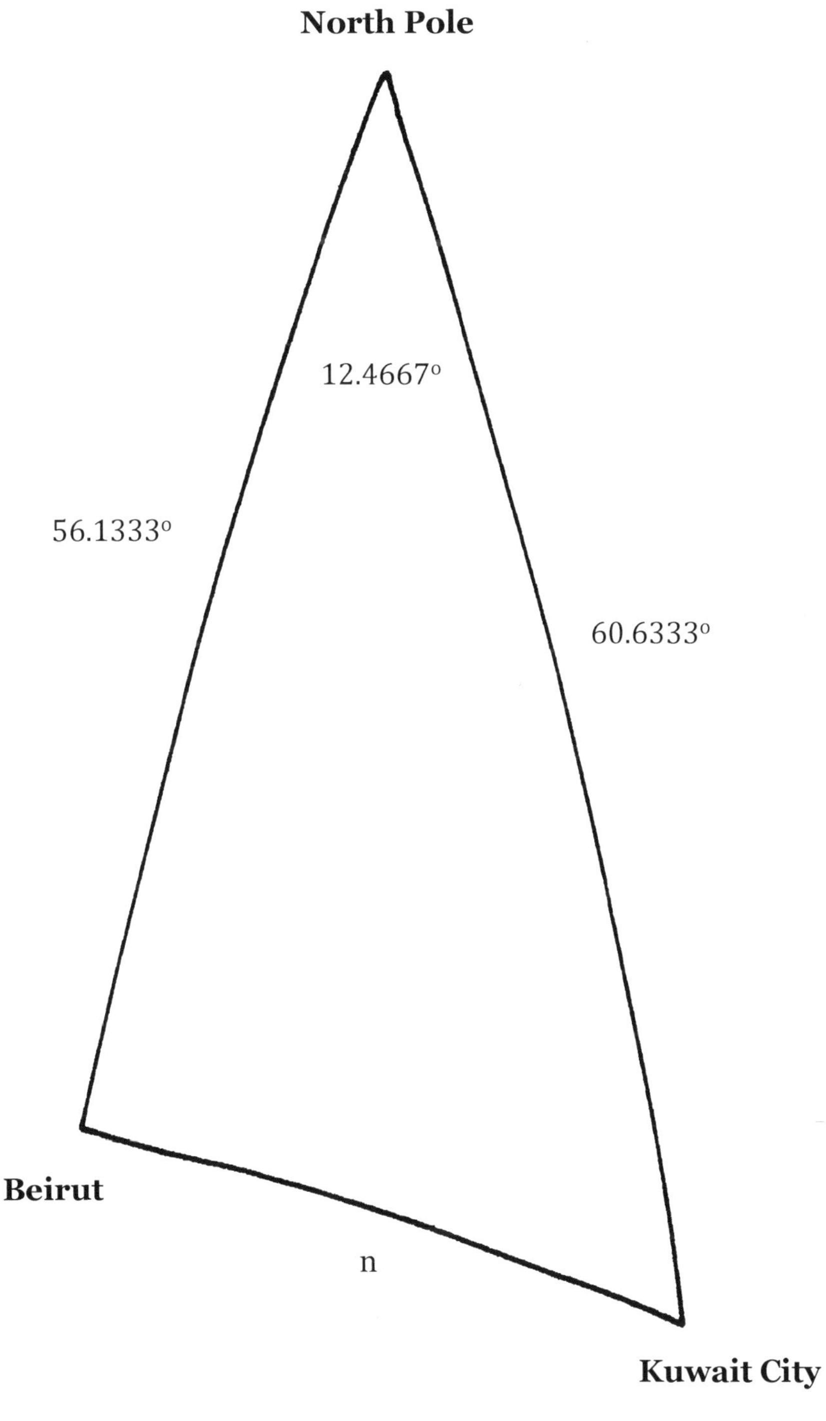
North Pole
12.4667°
56.1333°
60.6333°
Beirut
n
Kuwait City

Chapter 7
Problem 5
Solution

Damascus, Syria is located at $33^o30'$ N $= \left(33 + {}^{30}/_{60}\right)^o$ N $= 33.5000^o$ N ,

$36^o18'$ E $= \left(36 + {}^{18}/_{60}\right)^o$ E $= 36.3000^o$ E .

Doha, Qatar is located at $25^o17'$ N $= \left(25 + {}^{17}/_{60}\right)^o$ N $\approx 25.2833^o$ N ,

$51^o32'$ E $= \left(51 + {}^{32}/_{60}\right)^o$ E $\approx 51.5333^o$ E .

Two sides of the spherical triangle are $90^o - 33.5000^o = 56.5000^o$ and $90^o - 25.2833^o = 64.7167^o$, respectively. The angle at the North Pole is $51.5333^o - 36.3000^o = 15.2333^o$.

See the illustration on the next page.

It follows that

$\cos n = \cos(56.5000^o)\cos(64.7167^o) + \sin(56.5000^o)\ \sin(64.7167^o)\cos(15.2333^o)$

$\cos n \approx 0.963241842226 \ \Rightarrow\ n \approx \cos^{-1}(0.963241842226)\ \Rightarrow\ n \approx 15.5831007328^o$.

Therefore, the approximate distance between the two capital cities is

$(15.5831007328^o)\left({}^{\pi}/_{180^o}\right)(3959 \text{ miles})$

$\approx (0.271976415457)(3959 \text{ miles})$

$\approx$ 1,077 miles.

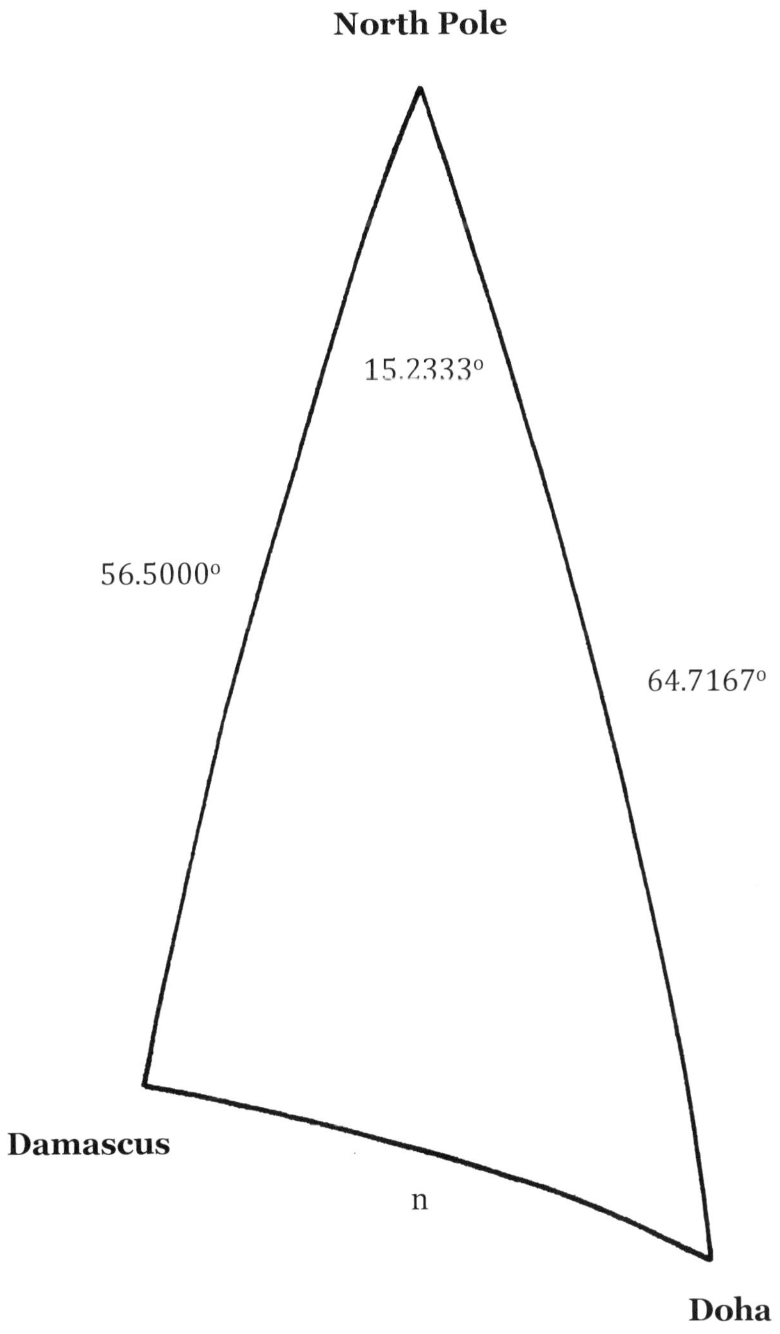
North Pole
15.2333°
56.5000°
64.7167°
Damascus
n
Doha

Chapter 7
Problem 6
Solution

Abu Dhabi, United Arab Emirates is located at

$24^o28'$ N $= \left(24 + {}^{28}/_{60}\right)^o$ N $\approx 24.4667^o$ N , $54^o22'$ E $= \left(54 + {}^{22}/_{60}\right)^o$ E $\approx 54.3667^o$ E .

Muscat, Oman is located at $23^o37'$ N $= \left(23 + {}^{37}/_{60}\right)^o$ N $\approx 23.6167^o$ N ,

$58^o35'$ E $= \left(58 + {}^{35}/_{60}\right)^o$ E $\approx 58.5833^o$ E .

Two sides of the spherical triangle are $90^o - 24.4667^o = 65.5333^o$ and $90^o - 23.6167^o = 66.3833^o$, respectively. The angle at the North Pole is $58.5833^o - 54.3667^o = 4.2166^o$.

See the illustration on the next page.

It follows that

$\cos n = \cos(65.5333^o)\cos(66.3833^o) + \sin(65.5333^o)\ \sin(66.3833^o)\cos(4.2166^o)$

$\cos n \approx 0.997632585231 \ \Rightarrow\ n \approx \cos^{-1}(0.997632585231) \ \Rightarrow\ n \approx 3.94330639722^o$

Therefore, the approximate distance between the two state capital cities is

$(3.94330639722^o)\left({}^{\pi}/_{180^o}\right)(3959 \text{ miles})$

$\approx (0.068823680047)(3959 \text{ miles})$

≈ 272 miles.

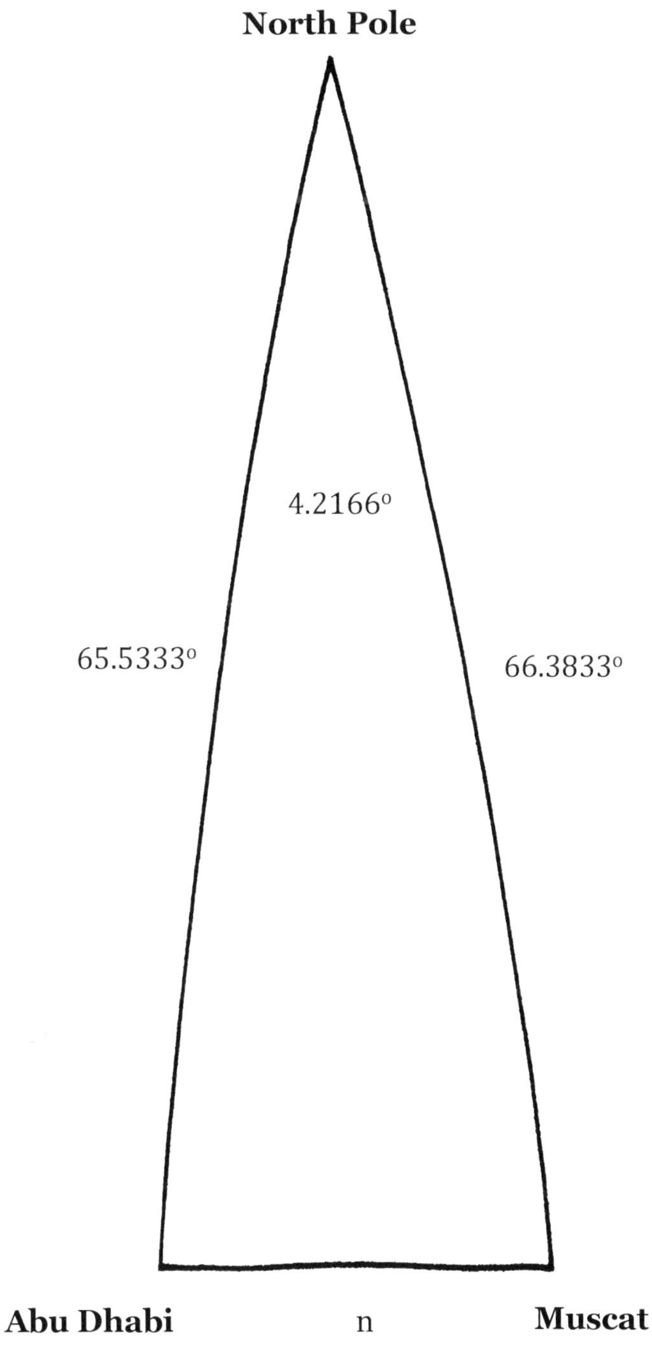
North Pole
4.2166°
65.5333°
66.3833°
Abu Dhabi
n
Muscat

Chapter 7
Problem 7
Solution

Ankara, Turkey is located at $39^o 56' \text{ N} = \left(39 + {}^{56}/_{60}\right)^o \text{ N} \approx 39.9333^o \text{ N}$,

$32^o 52' \text{ E} = \left(32 + {}^{52}/_{60}\right)^o \text{ E} \approx 32.8667^o \text{ E}$.

Baghdad, Iraq is located at $33^o 20' \text{ N} = \left(33 + {}^{20}/_{60}\right)^o \text{ N} \approx 33.3333^o \text{ N}$,

$44^o 24' \text{ E} = \left(44 + {}^{24}/_{60}\right)^o \text{ E} \approx 44.4000^o \text{ E}$.

Two sides of the spherical triangle are $90^o - 39.9333^o = 50.0667^o$ and $90^o - 33.3333^o = 56.6667^o$, respectively. The angle at the North Pole is $44.4000^o - 32.8667^o = 11.5333^o$.

See the illustration on the next page.

It follows that

$$\cos n = \cos(50.0667^o)\cos(56.6667^o) + \sin(50.0667^o)\ \sin(56.6667^o)\cos(11.5333^o)$$

$$\cos n \approx 0.9804372734 \quad \Rightarrow \; n \approx \cos^{-1}(0.9804372734) \;\Rightarrow\; n \approx 11.35175149^o.$$

Therefore, the approximate distance between the two capital cities is

$$(11.35175149^o)\left({}^{\pi}/_{180^o}\right)(3959 \text{ miles})$$

$$\approx (0.1981254393)(3959 \text{ miles})$$

$$\approx\ 784\ \text{ miles}.$$

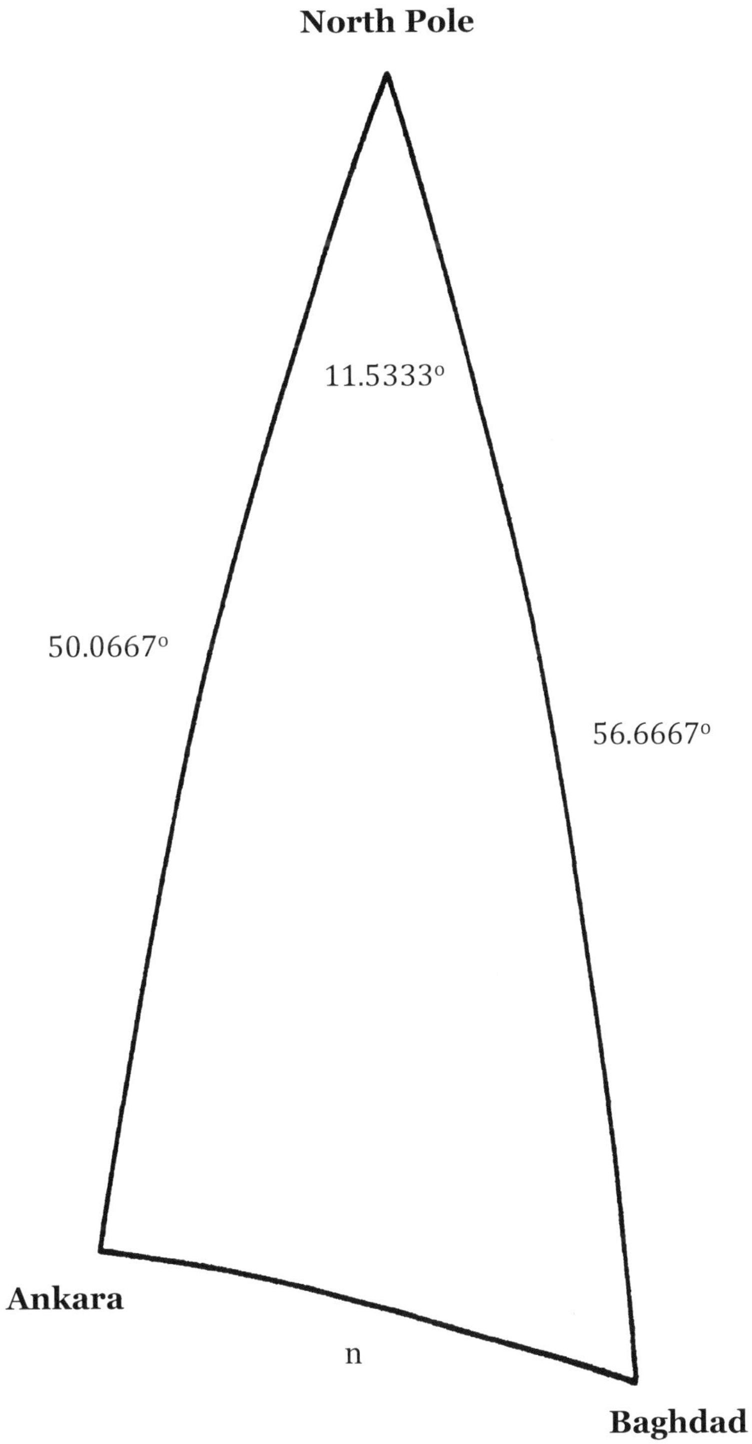
North Pole
11.5333°
50.0667°
56.6667°
Ankara
n
Baghdad

Chapter 7
Problem 8
Solution

Sanaa, Yemen is located at $15^o21'$ N $= \left(15 + {}^{21}/_{60}\right)^o$ N $= 15.3500^o$ N,

$44^o12'$ E $= \left(44 + {}^{12}/_{60}\right)^o$ E $= 44.2000^o$ E.

Tehran, Iran is located at $35^o42'$ N $= \left(35 + {}^{42}/_{60}\right)^o$ N $= 35.7000^o$ N,

$51^o25'$ E $= \left(51 + {}^{25}/_{60}\right)^o$ E $\approx 51.4167^o$ E.

Two sides of the spherical triangle are $90^o - 15.3500^o = 74.6500^o$ and $90^o - 35.7000^o = 54.3000^o$, respectively. The angle at the North Pole is $51.4167^o - 44.2000^o = 7.2167^o$.

See the illustration on the next page.

It follows that

$\cos n = \cos(74.6500^o)\cos(54.3000^o) + \sin(74.6500^o)\ \sin(54.3000^o)\cos(7.2167^o)$

$\cos n \approx 0.931382091282 \ \Rightarrow\ n \approx \cos^{-1}(0.931382091282) \ \Rightarrow\ n \approx 21.3487071134^o$.

Therefore, the approximate distance between the two capital cities is

$(21.3487071134^o)\left({}^{\pi}/_{180^o}\right)(3959 \text{ miles})$

$\approx (0.372605230174)(3959 \text{ miles})$

$\approx\ 1{,}475$ miles.

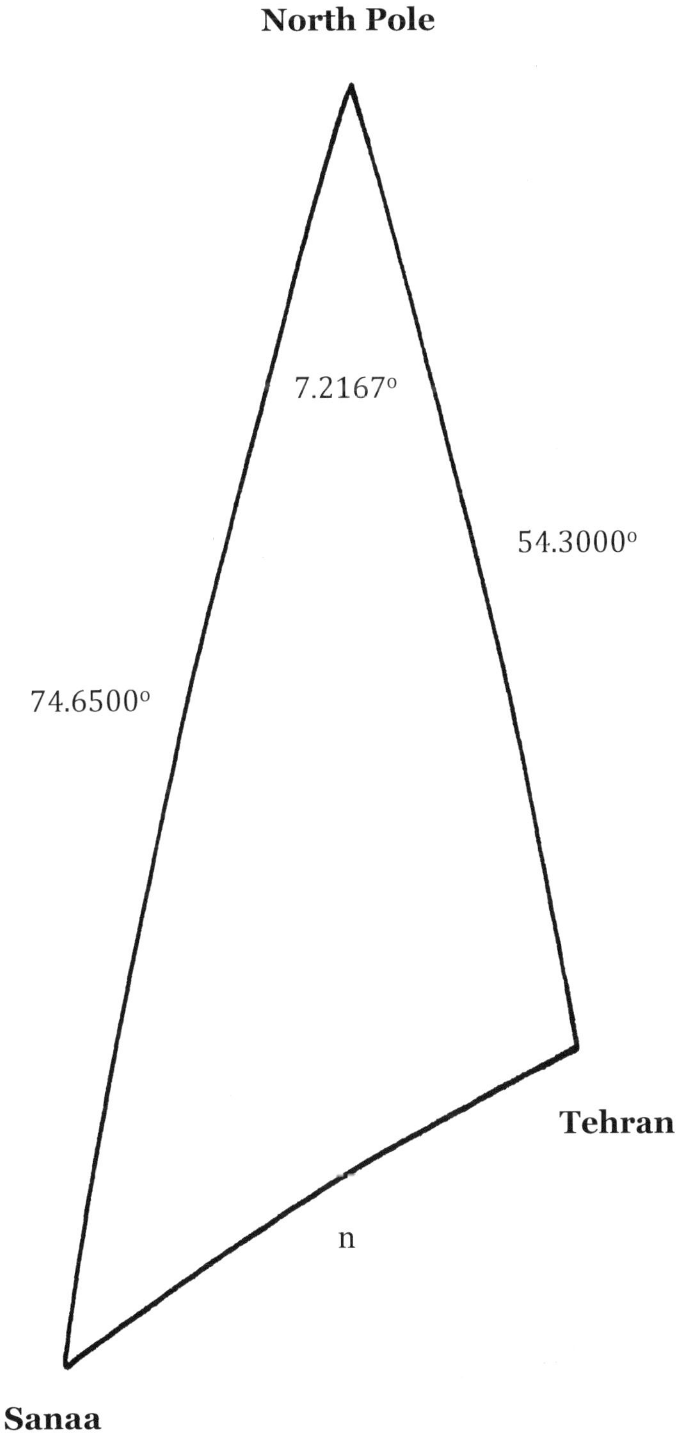
North Pole
7.2167°
54.3000°
74.6500°
Tehran
n
Sanaa

Chapter 8

North America

Example 1

Ottawa, Canada is located at $45^o25' \text{ N} = \left(45 + {}^{25}/_{60}\right)^o \text{ N} \approx 45.4167^o \text{ N}$,

$75^o42' \text{ W} = \left(75 + {}^{42}/_{60}\right)^o \text{ W} = 75.7000^o \text{ W}$.

Hamilton, Bermuda is located at $32^o17' \text{ N} = \left(32 + {}^{17}/_{60}\right)^o \text{ N} \approx 32.2833^o \text{ N}$,

$64^o47' \text{ W} = \left(64 + {}^{47}/_{60}\right)^o \text{ W} \approx 64.7833^o \text{ W}$.

Two sides of the spherical triangle are $90^o - 45.4167^o = 44.5833^o$ and $90^o - 32.2833^o = 57.7167^o$, respectively. The angle at the North Pole is $75.7000^o - 64.7833^o = 10.9167^o$.

See the illustration on the next page.

It follows that

$\cos n = \cos(44.5833^o)\cos(57.7167^o) + \sin(44.5833^o)\ \sin(57.7167^o)\cos(10.9167^o)$

$\cos n \approx 0.963104577284 \ \Rightarrow\ n \approx \cos^{-1}(0.963104577284) \ \Rightarrow\ n \approx 15.6123504186^o$.

Therefore, the approximate distance between the two capital cities is

$(15.6123504186^o)\left({}^{\pi}/_{180^o}\right)(3959 \text{ miles})$

$\approx (0.272486918779)(3959 \text{ miles})$

$\approx\ 1{,}079\ \text{ miles}$.

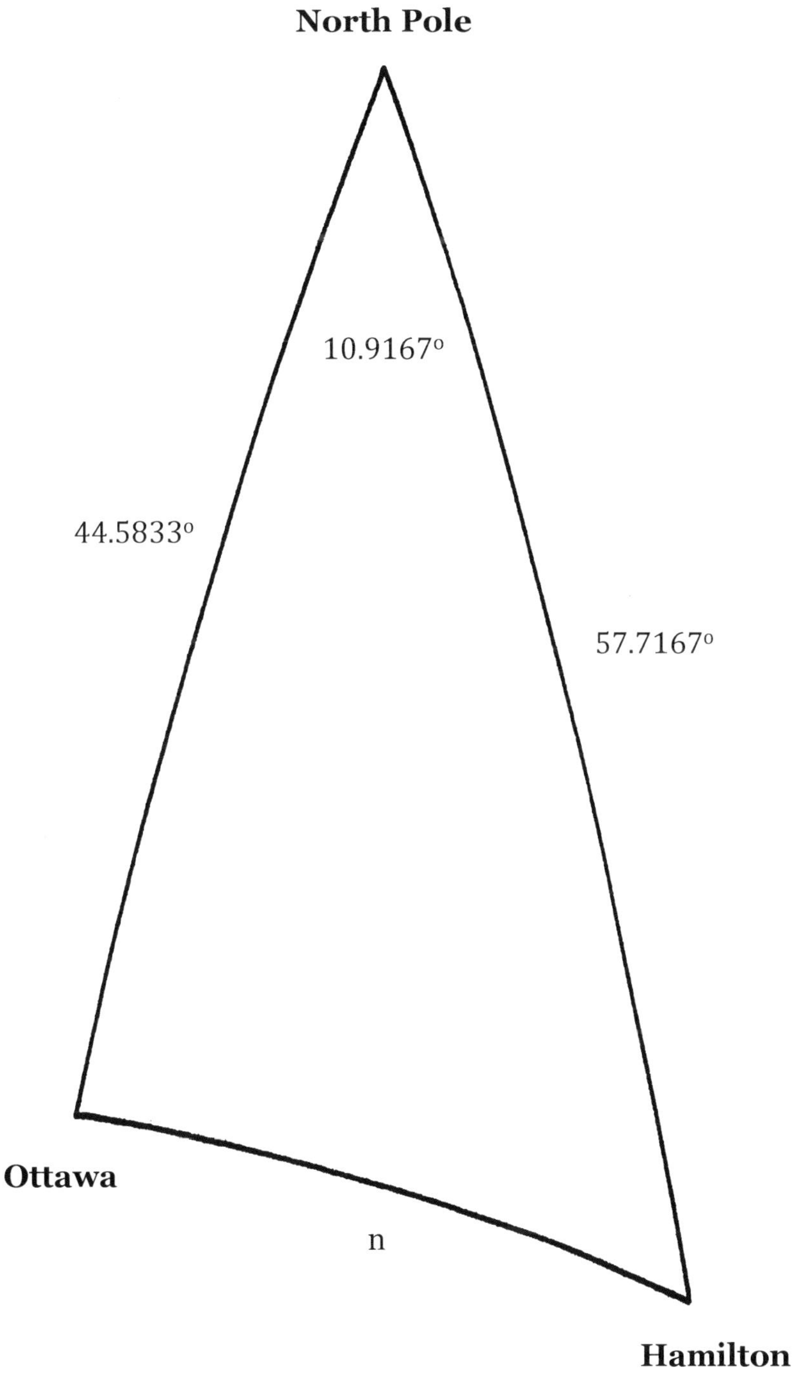
North Pole
10.9167°
44.5833°
57.7167°
Ottawa
n
Hamilton

Chapter 8 Problems:

Approximate the distance between the following capitals in North America.

1. Mexico City, Mexico ($19^\circ 26'$ N $99^\circ 8'$ W)
   to Washington, D.C., United States ($38^\circ 53'$ N $77^\circ 2'$ W)

2. Saint-Pierre, Saint Pierre and Miquelon ($46^\circ 46'$ N $56^\circ 11'$ W)
   to Nuuk, Greenland ($64^\circ 11'$ N $51^\circ 45'$ W)

Chapter 8

Problem 1

Solution

Mexico City, Mexico is located at $19^o 26'$ N $= \left(19 + {}^{26}/_{60}\right)^o$ N $\approx 19.4333^o$ N ,

$99^o 8'$ W $= \left(99 + {}^{8}/_{60}\right)^o$ W $\approx 99.1333^o$ W .

Washington, D.C., United States is located at

$38^o 53'$ N $= \left(38 + {}^{53}/_{60}\right)^o$ N $\approx 38.8833^o$ N , $77^o 2'$ W $= \left(77 + {}^{2}/_{60}\right)^o$ W $\approx 77.0333^o$ W.

Two sides of the spherical triangle are $90^o - 19.4333^o = 70.5667^o$ and $90^o - 38.8833^o = 51.1167^o$, respectively. The angle at the North Pole is $99.1333^o - 77.0333^o = 22.1000^o$.

See the illustration on the next page.

It follows that

$$\cos n = \cos(70.5667^o)\cos(51.1167^o) + \sin(70.5667^o)\ \sin(51.1167^o)\cos(22.1000^o)$$

$$\cos n \approx 0.888998660488 \ \Rightarrow\ n \approx \cos^{-1}(0.888998660488) \ \Rightarrow\ n \approx 27.2523125181^o.$$

Therefore, the approximate distance between the two capital cities is

$(27.2523125181^o)\left({}^{\pi}/_{180^o}\right)(3959 \text{ miles})$

$\approx (0.475642582223)(3959 \text{ miles})$

$\approx$ 1,883 miles.

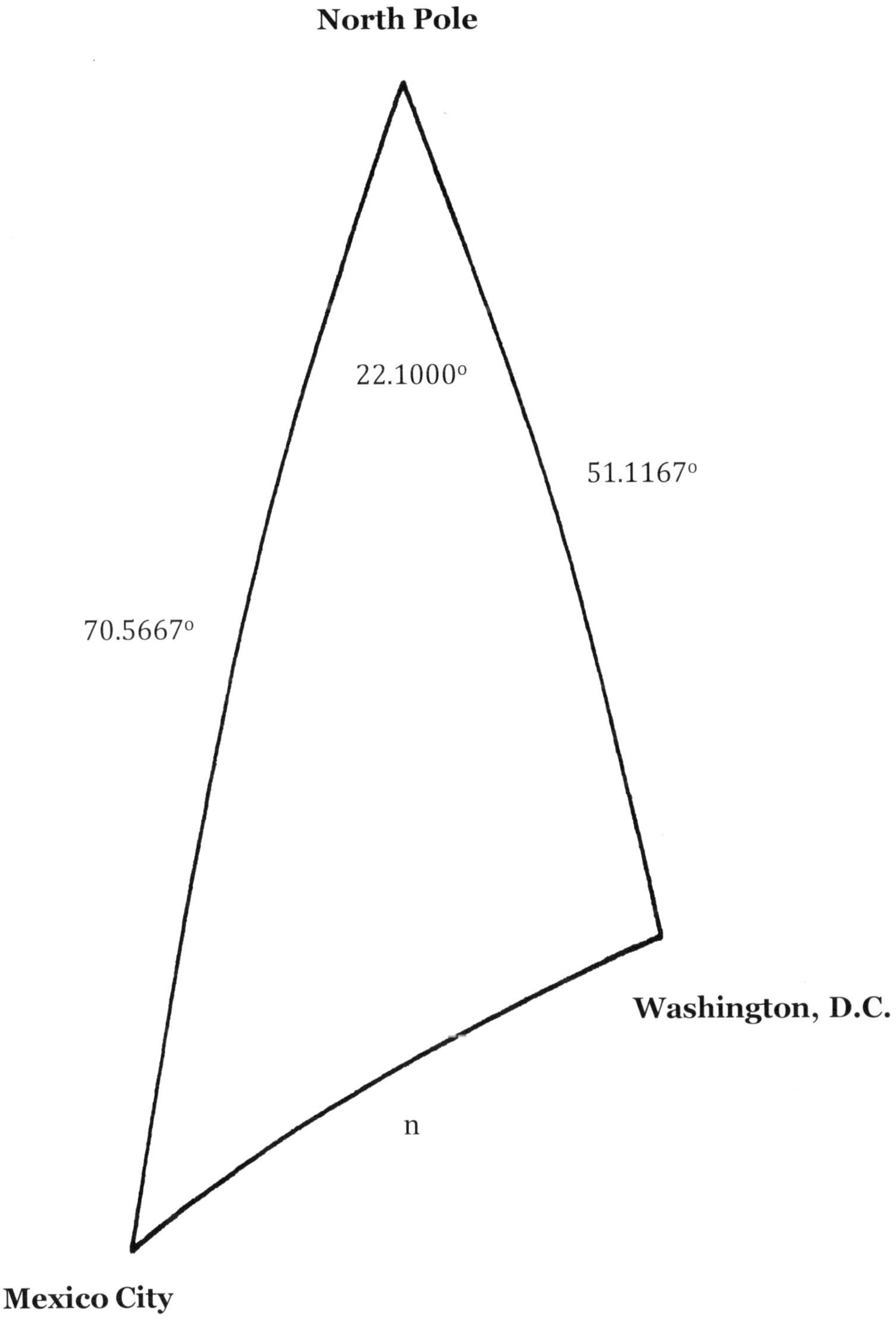
North Pole
22.1000°
51.1167°
70.5667°
Washington, D.C.
n
Mexico City

Chapter 8
Problem 2
Solution

Saint-Pierre, Saint Pierre and Miquelon is located at

$46^o46'\text{ N} = \left(46 + {}^{46}/_{60}\right)^o \text{ N} \approx 46.7667^o \text{ N}$,

$56^o11'\text{ W} = \left(56 + {}^{11}/_{60}\right)^o \text{ W} \approx 56.1833^o \text{ W}$.

Nuuk, Greenland is located at $64^o11'\text{ N} = \left(64 + {}^{11}/_{60}\right)^o \text{ N} \approx 64.1833^o \text{ N}$,

$51^o45'\text{ W} = \left(51 + {}^{45}/_{60}\right)^o \text{ W} = 51.7500^o \text{ W}$.

Two sides of the spherical triangle are $90^o - 46.7667^o = 43.2333^o$ and $90^o - 64.1833^o = 25.8167^o$, respectively. The angle at the North Pole is $56.1833^o - 51.7500^o = 4.4333^o$.

See the illustration on the next page.

It follows that

$\cos n = \cos(43.2333^o)\cos(25.8167^o) + \sin(43.2333^o)\ \sin(25.8167^o)\cos(4.4333^o)$

$\cos n \approx 0.953261133124 \Rightarrow n \approx \cos^{-1}(0.953261133124\ ) \Rightarrow n \approx 17.5866426839^o$.

Therefore, the approximate distance between the two capital cities is

$(17.5866426839^o)\left({}^{\pi}/_{180^o}\right)(3959 \text{ miles})$

$\approx (0.306944819206)(3959 \text{ miles})$

$\approx\ 1{,}215\ \text{ miles}$.

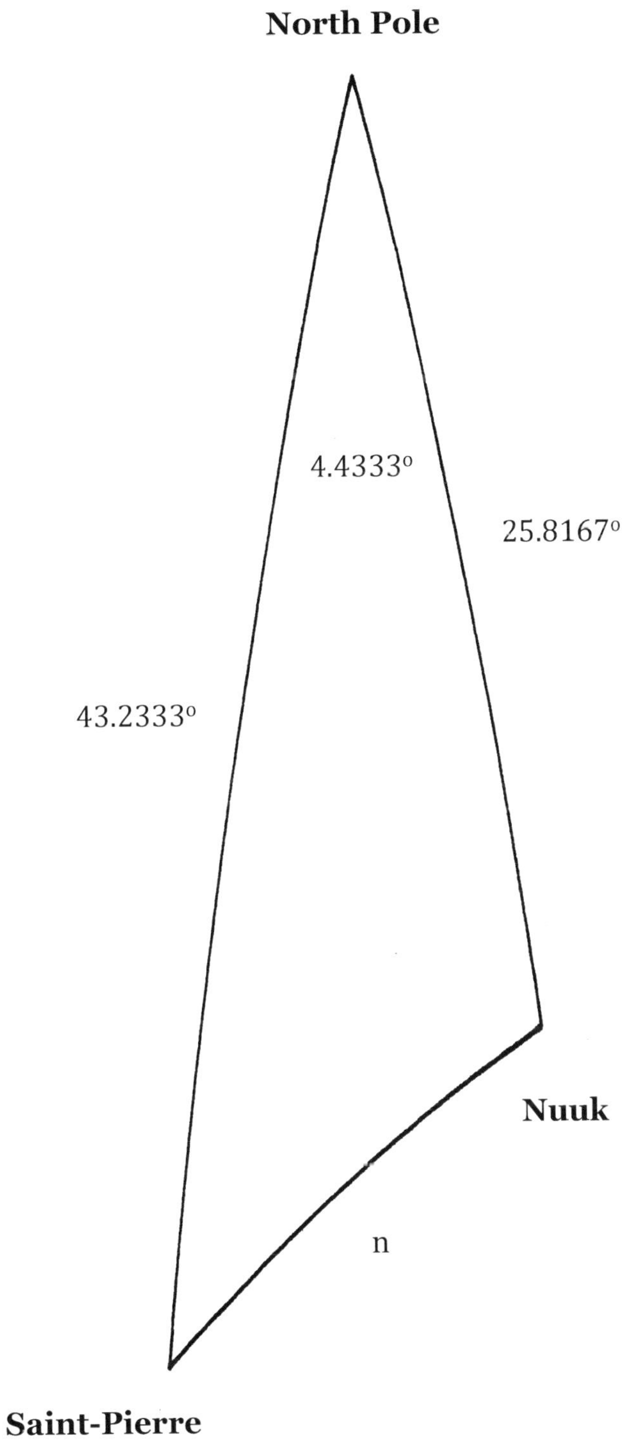
North Pole
4.4333°
25.8167°
43.2333°
Nuuk
n
Saint-Pierre

Chapter 9

South America

Example 1

Quito, Ecuador is located at $0^o13' \text{ S} = \left({}^{13}/_{60}\right)^o \text{ S} \approx 0.2167^o \text{ S}$,

$78^o30' \text{ W} = \left(78 + {}^{30}/_{60}\right)^o \text{ W} = 78.5000^o \text{ W}$.

Santiago, Chile is located at $33^o27' \text{ S} = \left(33 + {}^{27}/_{60}\right)^o \text{ S} = 33.4500^o \text{ S}$,

$70^o40' \text{ W} = \left(70 + {}^{40}/_{60}\right)^o \text{ W} \approx 70.6667^o \text{ W}$.

Two sides of the spherical triangle are $90^o - 0.2167^o = 89.7833^o$ and $90^o - 33.4500^o = 56.5500^o$, respectively. The angle at the South Pole is $78.5000^o - 70.6667^o = 7.8333^o$.

See the illustration on the next page.

It follows that

$\cos n = \cos(89.7833^o)\cos(56.5500^o) + \sin(89.7833^o)\ \sin(56.5500^o)\cos(7.8333^o)$

$\cos n \approx\ 0.828660336827 \Rightarrow\ n \approx \cos^{-1}(0.828660336827)\ \Rightarrow\ n \approx 34.0386328561^o$.

Therefore, the approximate distance between the two capital cities is

$(34.0386328561^o)\left({}^{\pi}/_{180^o}\right)(3959 \text{ miles})$

$\approx\ (0.594086216216)(3959 \text{ miles})$

$\approx\ 2{,}352$ miles.

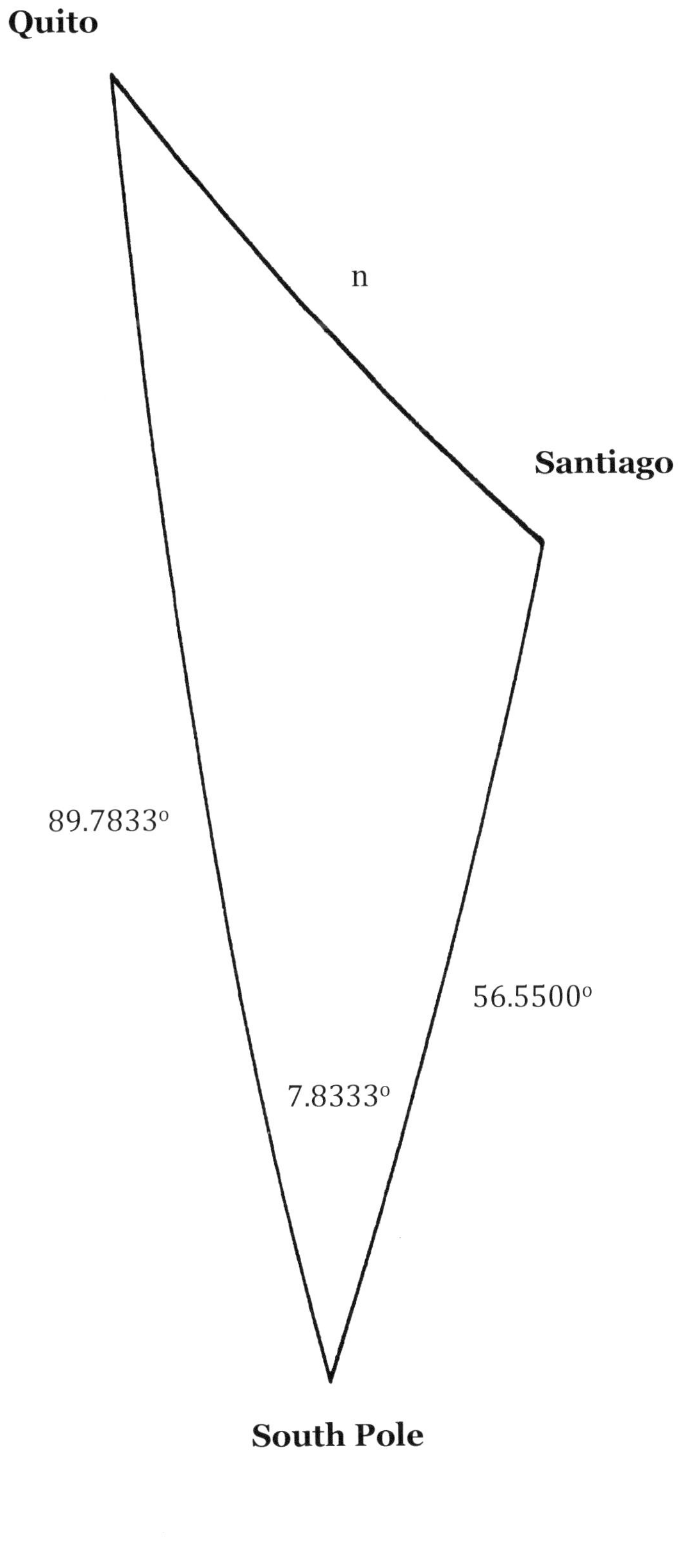
Quito
n
Santiago
89.7833°
56.5500°
7.8333°
South Pole

Example 2

Lima, Peru is located at $12^o3' \text{ S} = \left(12 + {}^{3}/_{60}\right)^o \text{ S} = 12.0500^o \text{ S}$,

$77^o3' \text{ W} = \left(77 + {}^{3}/_{60}\right)^o \text{ W} = 77.0500^o \text{ W}$.

Caracas, Venezuela is located at $10^o29' \text{ N} = \left(10 + {}^{29}/_{60}\right)^o \text{ N} \approx 10.4833^o \text{ N}$,

$66^o52' \text{ W} = \left(66 + {}^{52}/_{60}\right)^o \text{ W} \approx 66.8667^o \text{ W}$.

Two sides of the spherical triangles are $90^o + 12.0500^o = 102.0500^o$ and $90^o - 10.4833^o = 79.5167^o$, respectively. The angle at the North Pole is $77.0500^o - 66.8667^o = 10.1833^o$.

See the illustration on the next page.

It follows that

$\cos n = \cos(102.0500^o)\cos(79.5167^o) + \sin(102.0500^o)\sin(79.5167^o)\cos(10.1833^o)$

$\cos n = 0.908508397402 \approx \;\Rightarrow n \approx \cos^{-1}(0.908508397402) \;\Rightarrow n \approx 24.6999698518^o$.

Therefore, the approximate distance between the two capital cities

$(24.6999698518^o)\left({}^{\pi}/_{180^o}\right)(3959 \text{ miles})$

$\approx (0.431095799058)(3959 \text{ miles})$

$\approx$ 1,707 miles.

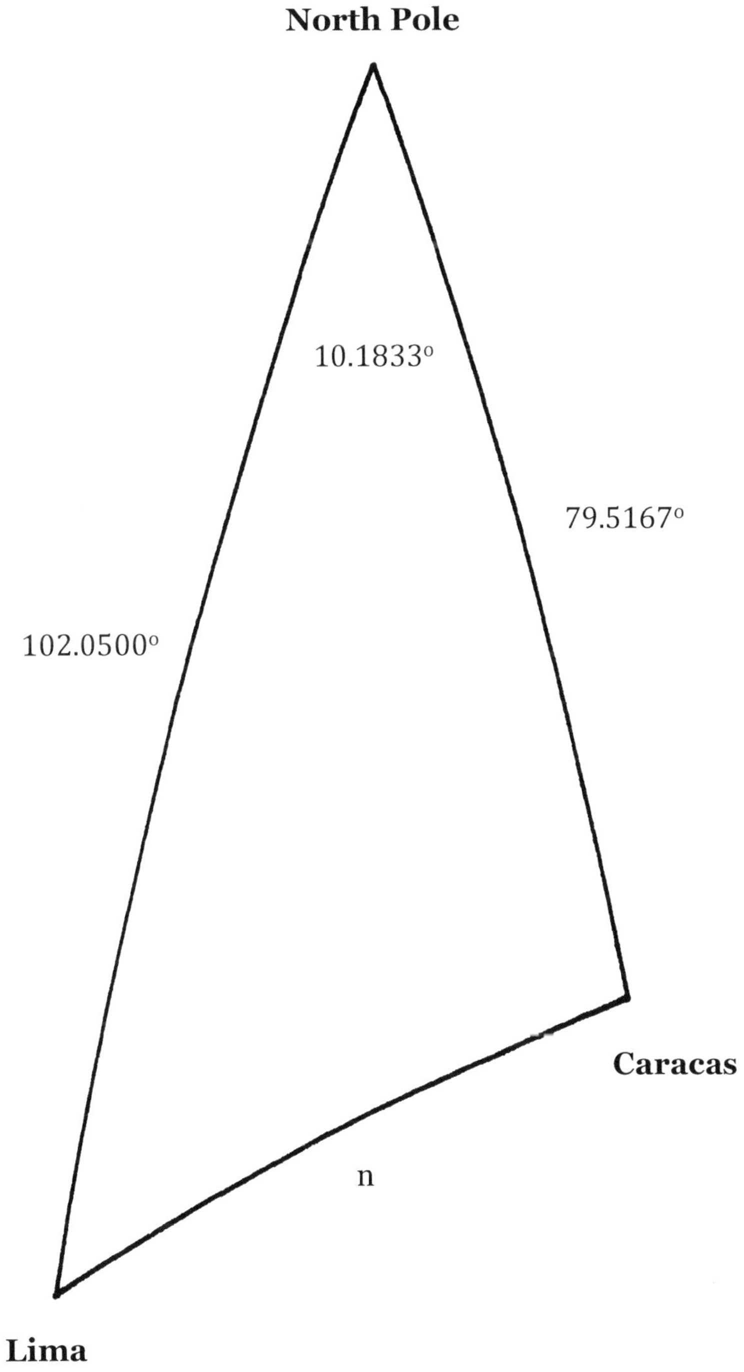
North Pole
10.1833°
79.5167°
102.0500°
Caracas
n
Lima

Chapter 9 Problems:

Approximate the distance between the following capitals in South America.

1. La Paz, Bolivia ($16°30'$ S $68°9'$ W)
   to Buenos Aires, Argentina ($34°36'$ S $58°22'$ W)

2. Bogota, Colombia ($4°36'$ N $74°5'$ W)
   to Paramaribo, Suriname ($5°50'$ N $55°10'$ W)

3. Stanley, Falkland Islands ($51°42'$ S $57°51'$ W)
   to Montevideo, Uruguay ($34°51'$ S $56°10'$ W)

4. Georgetown, Guyana ($6°48'$ N $58°9'$ W)
   to Paramaribo, Suriname ($5°50'$ N $55°10'$ W)

5. Asuncion, Paraguay ($25°16'$ S $57°40'$ W)
   to Brasilia, Brazil ($15°47'$ S $47°55'$ W)

Chapter 9
Problem 1
Solution

La Paz, Bolivia is located at $16^o30' \text{ S} = \left(16 + {}^{30}/_{60}\right)^o \text{ S} = 16.5000^o \text{ S}$,

$68^o9' \text{ W} = \left(68 + {}^{9}/_{60}\right)^o \text{ W} = 68.1500^o \text{ W}$.

Buenos Aires, Argentina is located at $34^o36' \text{ S} = \left(34 + {}^{36}/_{60}\right)^o \text{ S} \approx 34.6000^o \text{ S}$,

$58^o22' \text{ W} = \left(58 + {}^{22}/_{60}\right)^o \text{ W} \approx 58.3667^o \text{ W}$.

Two sides of the spherical triangle are $90^o - 16.5000^o = 73.5000^o$ and $90^o - 34.6000^o = 55.4000^o$, respectively. The angle at the South Pole is $68.1500^o - 58.3667^o = 9.7833^o$.

See the illustration on the next page.

It follows that

$\cos n = \cos(73.5000^o)\cos(55.4000^o) + \sin(73.5000^o)\ \sin(55.4000^o)\cos(9.7833^o)$

$\cos n \approx 0.9390381921 \ \Rightarrow n \approx \cos^{-1}(0.9390381921) \ \Rightarrow n \approx 20.10934437^o$.

Therefore, the approximate distance between the two capital cities

$(20.10934437^o)\left({}^{\pi}/_{180^o}\right)(3959 \text{ miles})$

$\approx (0.3509742696)(3959 \text{ miles})$

$\approx$ 1,390 miles.

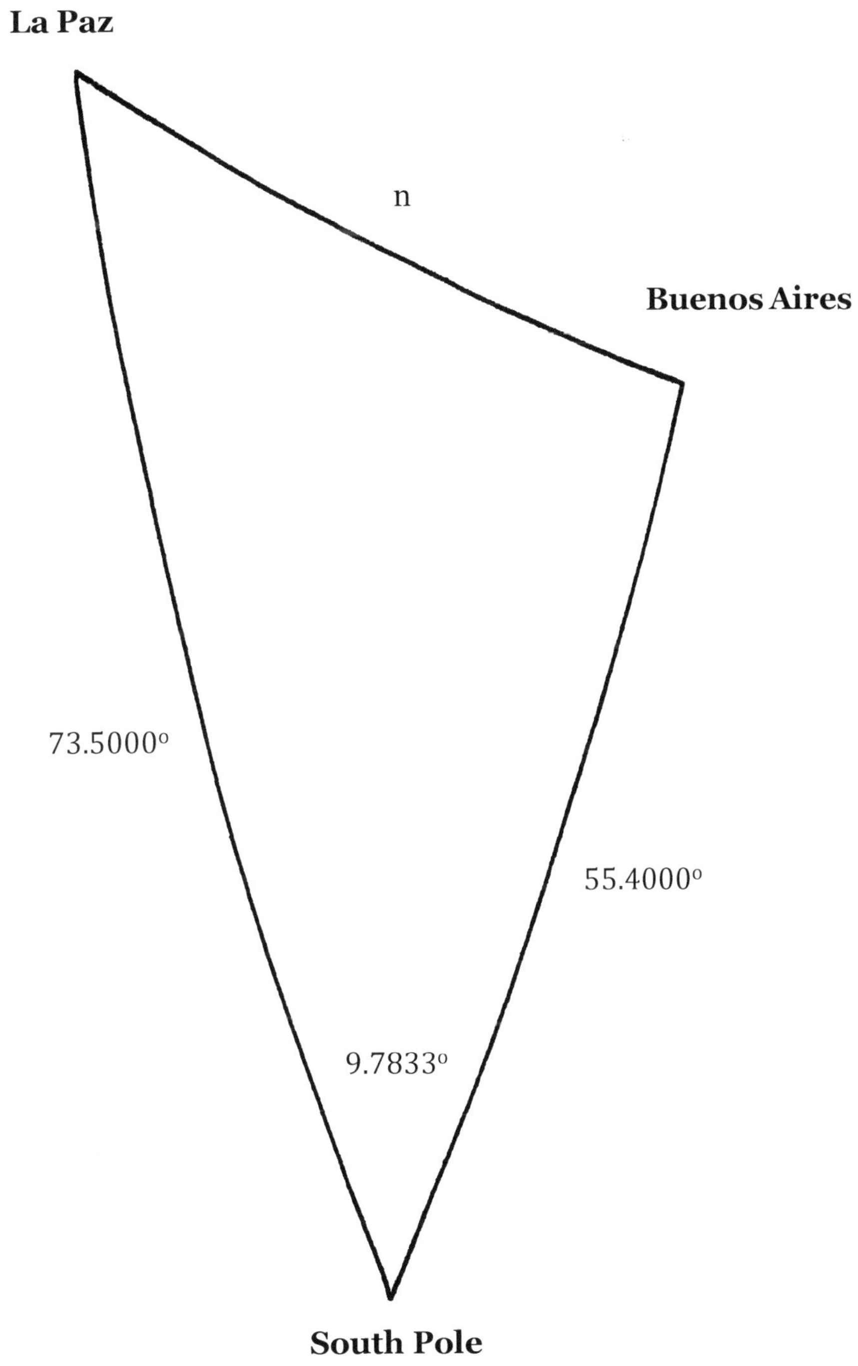
La Paz
n
Buenos Aires
73.5000°
55.4000°
9.7833°
South Pole

Chapter 9

Problem 2

Solution

Bogota, Colombia is located at $4^o 36' \text{ N} = \left(4 + {}^{36}/_{60}\right)^o \text{ N} = 4.6000^o \text{ N}$,

$74^o 5' \text{ W} = \left(74 + {}^{5}/_{60}\right)^o \text{ W} \approx 74.0833^o \text{ W}$.

Paramaribo, Suriname is located at $5^o 50' \text{ N} = \left(5 + {}^{50}/_{60}\right)^o \text{ N} \approx 5.8333^o \text{ N}$,

$55^o 10' \text{ W} = \left(55 + {}^{10}/_{60}\right)^o \text{ W} \approx 55.1667^o \text{ W}$.

Two sides of the spherical triangle are $90^o - 4.6000^o = 85.4000^o$ and $90^o - 5.8333^o = 84.1667^o$, respectively. The angle at the North Pole is $74.0833^o - 55.1667^o = 18.9166^o$.

See the illustration on the next page.

It follows that

$\cos \text{n} = \cos(85.4000^o)\cos(84.1667^o) + \sin(85.4000^o)\ \sin(84.1667^o)\cos(18.9166^o)$

$\cos \text{n} \approx 0.946212549005 \ \Rightarrow \text{n} \approx \cos^{-1}(0.946212549005) \ \Rightarrow \text{n} \approx 18.8774891875^o$.

Therefore, the approximate distance between the two capital cities

$(18.8774891875^o)\left({}^{\pi}/_{180^o}\right)(3959 \text{ miles})$

$\approx (0.329474340831)(3959 \text{ miles})$

$\approx$ 1,304 miles.

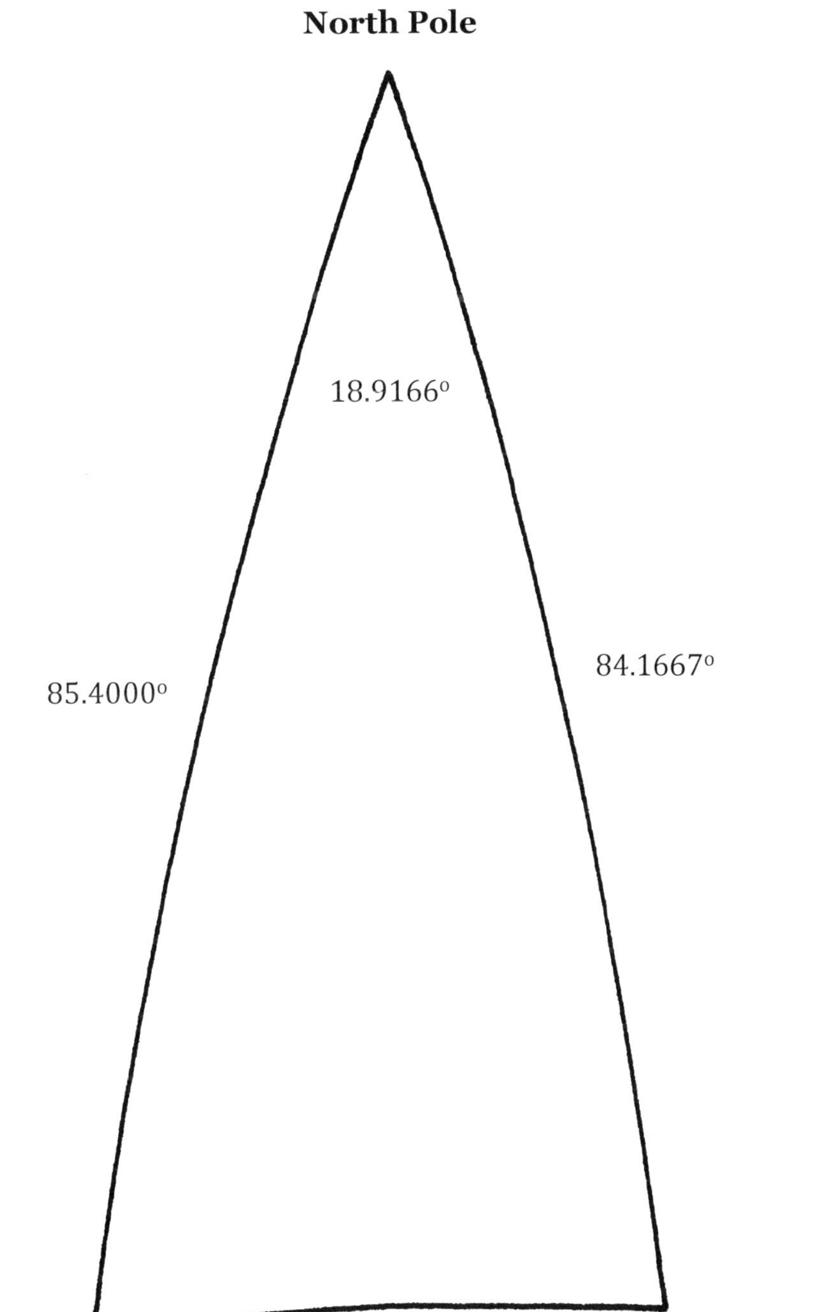
North Pole
18.9166°
85.4000°
84.1667°
n
Bogota
Paramaribo

Chapter 9

Problem 3

Solution

Stanley, Falkland Islands is located at $51^o42'\ S = \left(51 + {}^{42}/_{60}\right)^o\ S = 51.7000^o\ S$, $57^o51'\ W = \left(57 + {}^{51}/_{60}\right)^o\ W = 57.8500^o\ W$.

Montevideo, Uruguay is located at $34^o51'\ S = \left(34 + {}^{51}/_{60}\right)^o\ S = 34.8500^o\ S$, $56^o10'\ W = \left(56 + {}^{10}/_{60}\right)^o\ W \approx 56.1667^o\ W$.

Two sides of the spherical triangle are $90^o - 51.7000^o = 38.3000^o$ and $90^o - 34.8500^o = 55.1500^o$, respectively. The angle at the South Pole is $57.8500^o - 56.1667^o = 1.6833^o$.

See the illustration on the next page.

It follows that

$\cos n = \cos(38.3000^o)\cos(55.1500^o) + \sin(38.3000^o)\ \sin(55.1500^o)\cos(1.6833^o)$

$\cos n \approx 0.956847416677 \ \Rightarrow n \approx \cos^{-1}(0.956847416677) \ \Rightarrow n \approx 16.893330502^o$.

Therefore, the approximate distance between the two capital cities

$(16.893330502^o)\left({}^{\pi}/_{180^o}\right)(3959 \text{ miles})$

$\approx (0.294844238887)(3959 \text{ miles})$

$\approx\ 1{,}167$ miles.

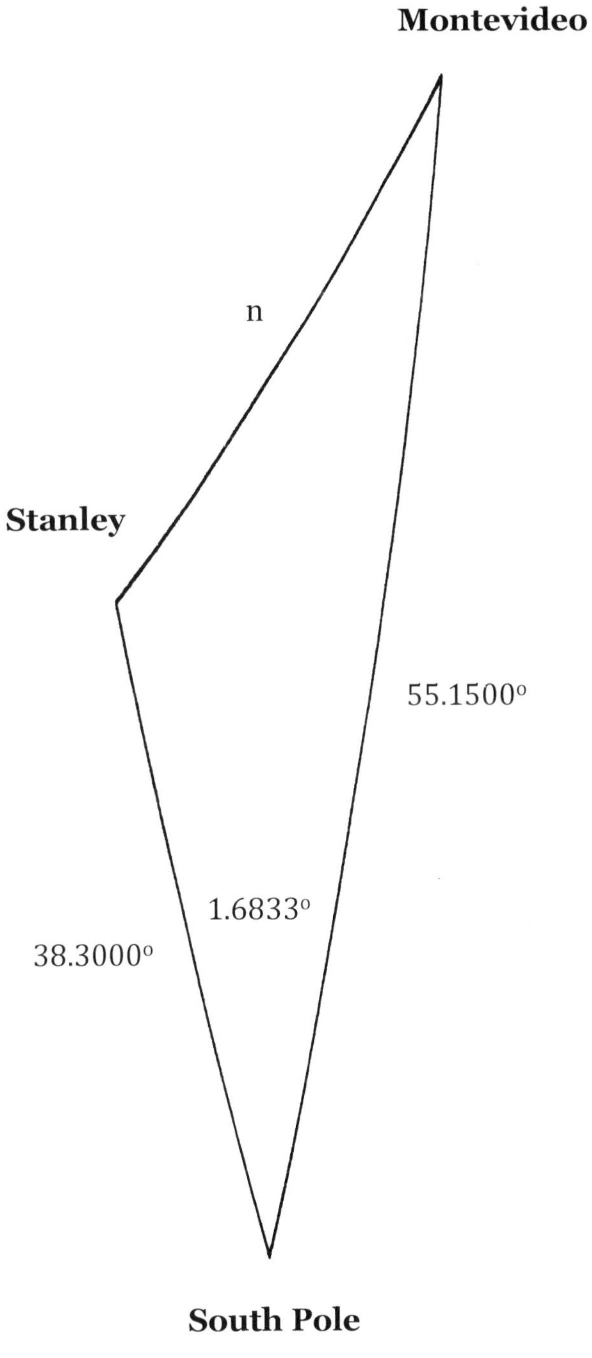

Montevideo
n
Stanley
55.1500°
1.6833°
38.3000°
South Pole

Chapter 9
Problem 4
Solution

Georgetown, Guyana is located at $6^o48'\ \mathrm{N} = \left(6 + {}^{48}/_{60}\right)^o\ \mathrm{N} = 6.8000^o\ \mathrm{N}$,

$58^o9'\ \mathrm{W} = \left(58 + {}^{9}/_{60}\right)^o\ \mathrm{W} = 58.1500^o\ \mathrm{W}$.

Paramaribo, Suriname is located at $5^o50'\ \mathrm{N} = \left(5 + {}^{50}/_{60}\right)^o\ \mathrm{N} \approx 5.8333^o\ \mathrm{N}$,

$55^o10'\ \mathrm{W} = \left(55 + {}^{10}/_{60}\right)^o\ \mathrm{W} \approx 55.1667^o\ \mathrm{W}$.

Two sides of the spherical triangle are $90^o - 6.8000^o = 83.2000^o$ and $90^o - 5.8333^o = 84.1667^o$, respectively. The angle at the North Pole is $58.1500^o - 55.1667^o = 2.9833^o$.

See the illustration on the next page.

It follows that

$$\cos n = \cos(83.2000^o)\cos(84.1667^o) + \sin(83.2000^o)\ \sin(84.1667^o)\cos(2.9833^o)$$

$$\cos n \approx 0.9985189181 \ \Rightarrow\ n \approx \cos^{-1}(0.9985189181\ ) \ \Rightarrow\ n \approx 3.118751543^o.$$

Therefore, the approximate distance between the two capital cities is

$(3.118751543^o)\left({}^{\pi}/_{180^o}\right)(3959\ \text{miles})$

$\approx (0.054432483)(3959\ \text{miles})$

$\approx\ 215\ \ \text{miles}$.

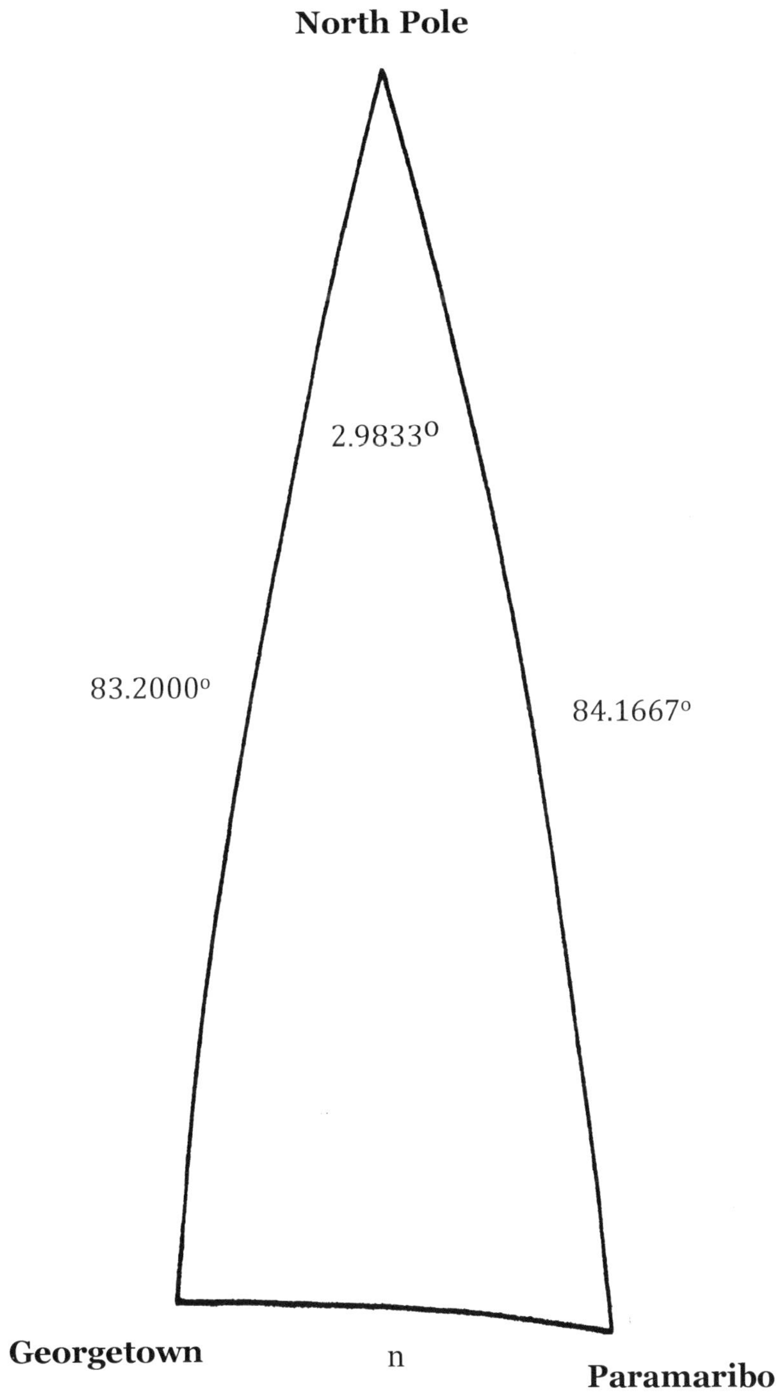

North Pole
2.9833°
83.2000°
84.1667°
Georgetown
n
Paramaribo

Chapter 9
Problem 5
Solution

Asuncion, Paraguay is located at $25^o16' \text{ S} = \left(25 + {}^{16}/_{60}\right)^o \text{ S} \approx 25.2667^o \text{ S}$,
$57^o40'\text{W} = \left(57 + {}^{40}/_{60}\right)^o \text{ W} \approx 57.6667^o \text{ W}$.

Brasilia, Brazil is located at $15^o47' \text{ S} = \left(15 + {}^{47}/_{60}\right)^o \text{ S} \approx 15.7833^o \text{ S}$,
$47^o55' \text{ W} = \left(47 + {}^{55}/_{60}\right)^o \text{ W} \approx 47.9167^o \text{ W}$.

Two sides of the spherical triangle are $90^o - 25.2667^o = 64.7333^o$ and $90^o - 15.7833^o = 74.2167^o$, respectively. The angle at the South Pole is $57.6667^o - 47.9167^o = 9.7500^o$.

See the illustration on the next page.

It follows that

$$\cos n = \cos(64.7333^o)\cos(74.2167^o) + \sin(64.7333^o)\ \sin(74.2167^o)\cos(9.7500^o)$$

$$\cos n \approx 0.97376375446 \Rightarrow n \approx \cos^{-1}(0.97376375446) \Rightarrow n \approx 13.1535383541^o.$$

Therefore, the approximate distance between the two capital cities is

$(13.1535383541^o)\left({}^{\pi}/_{180^o}\right)(3959 \text{ miles})$

$\approx (0.229572552566)(3959 \text{ miles})$

$\approx 909 \text{ miles}$

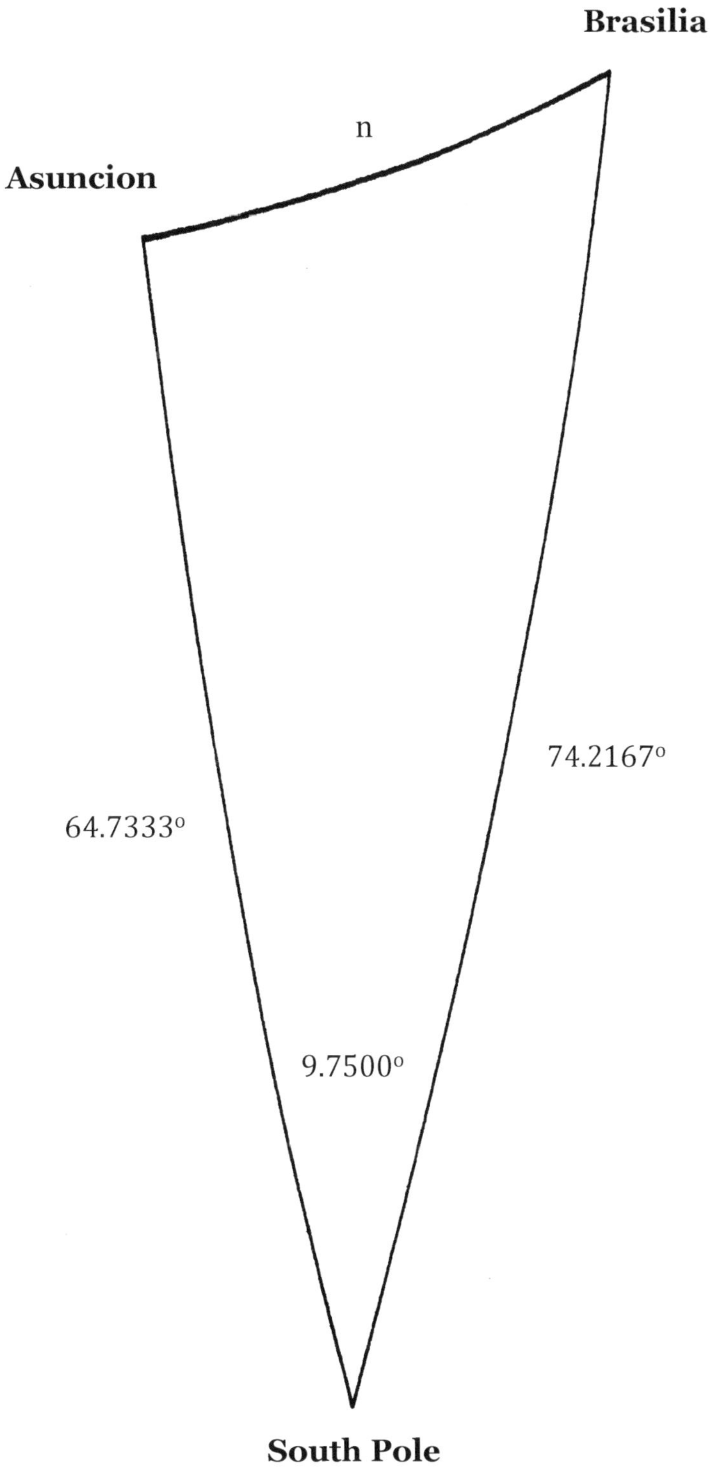
Brasilia
n
Asuncion
74.2167°
64.7333°
9.7500°
South Pole

Chapter 10

South Asia

Example 1

Colombo, Sri Lanka is located at $6^o 55' \text{ N} = \left(6 + {}^{55}/_{60}\right)^o \text{ N} \approx 6.9167^o \text{ N}$,

$79^o 50' \text{ E} = \left(79 + {}^{50}/_{60}\right)^o \text{ E} \approx 79.8333^o \text{ E}$.

Thimphu, Bhutan is located at $27^o 28' \text{ N} = \left(27 + {}^{28}/_{60}\right)^o \text{ N} \approx 27.4667^o \text{ N}$,

$89^o 38' \text{ E} = \left(89 + {}^{38}/_{60}\right)^o \text{ E} \approx 89.6333^o \text{ E}$.

Two sides of the spherical triangle are $90^o - 6.9167^o = 83.0833^o$ and $90^o - 27.4667^o = 62.5333^o$, respectively. The angle at the North Pole is $89.6333^o - 79.8333^o = 9.8000^o$.

See the illustration on the next page.

It follows that

$$\cos \text{n} = \cos(83.0833^o) \cos(62.5333^o) + \sin(83.0833^o) \sin(62.5333^o) \cos(9.8000^o)$$

$$\cos \text{n} \approx 0.923513179608 \Rightarrow \text{n} \approx \cos^{-1}(0.923513179608) \Rightarrow \text{n} \approx 22.5547875241^o.$$

Therefore, the approximate distance between the two capital cities is

$(22.5547875241^o)\left({}^{\pi}/_{180^o}\right)(3959 \text{ miles})$

$\approx (0.393655304383)(3959 \text{ miles})$

$\approx 1{,}558$ miles.

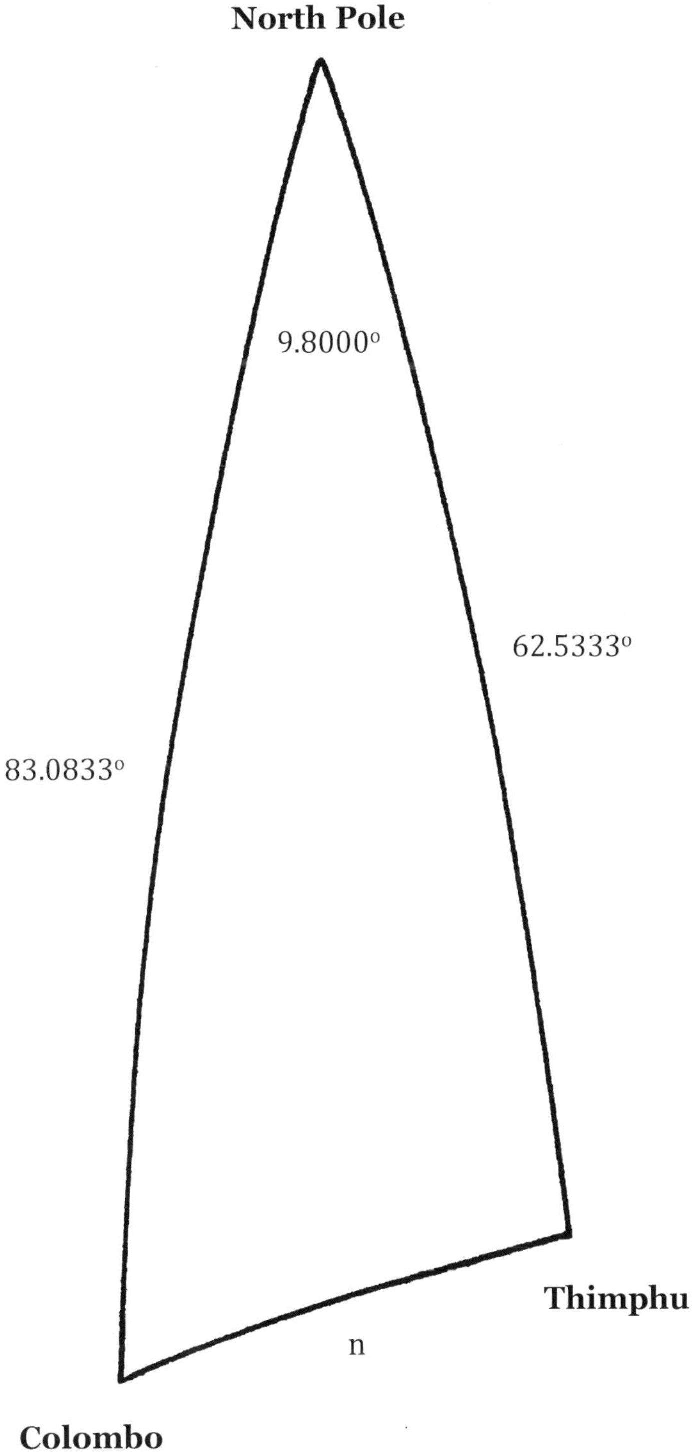
North Pole
9.8000°
62.5333°
83.0833°
Thimphu
n
Colombo

Chapter 10 Problems:

Approximate the distance between the following capitals in South Asia.

1. Kabul, Afghanistan $(34°31'N\ 69°11'\ E)$
   to Islamabad, Pakistan(33°41′N 73°3′ E)

2. Male, Maldives (4°10′N 73°30′ E)
   to Dhaka, Bangladesh (23°43′N 90°24′ E)

3. New Delhi, India (28°36′ N 77°12′ E)
   to Kathmandu, Nepal(27°43′ N 85°19′ E)

Chapter 10
Problem 1
Solution

Kabul, Afghanistan is located at $34^o31' \text{ N} = \left(34 + {}^{31}/_{60}\right)^o \text{ N} \approx 34.5167^o \text{ N}$, $69^o11' \text{ E} = \left(69 + {}^{11}/_{60}\right)^o \text{ E} \approx 69.1833^o \text{ E}$.

Islamabad, Pakistan is located at $33^o41' \text{ N} = \left(33 + {}^{41}/_{60}\right)^o \text{ N} \approx 33.6833^o \text{ N}$, $73^o3' \text{ E} = \left(73 + {}^{3}/_{60}\right)^o \text{ E} = 73.0500^o \text{ E}$.

Two sides of the spherical triangle are $90^o - 34.5167^o = 55.4833^o$ and $90^o - 33.6833^o = 56.3167^o$, respectively. The angle at the North Pole is $73.0500^o - 69.1833^o = 3.8667^o$.

See the illustration on the next page.

It follows that

$\cos n = \cos(55.4833^o)\cos(56.3167^o) + \sin(55.4833^o)\ \sin(56.3167^o)\cos(3.8667^o)$

$\cos n \approx 0.998333472274 \ \Rightarrow\ n \approx \cos^{-1}(0.998333472274) \ \Rightarrow\ n \approx 3.30829504173^o$.

Therefore, the approximate distance between the two capital cities is

$(3.30829504173^o)\left({}^{\pi}/_{180^o}\right)(3959 \text{ miles})$

$\approx (0.057740641106)(3959 \text{ miles})$

$\approx\ 229\ \text{ miles}$.

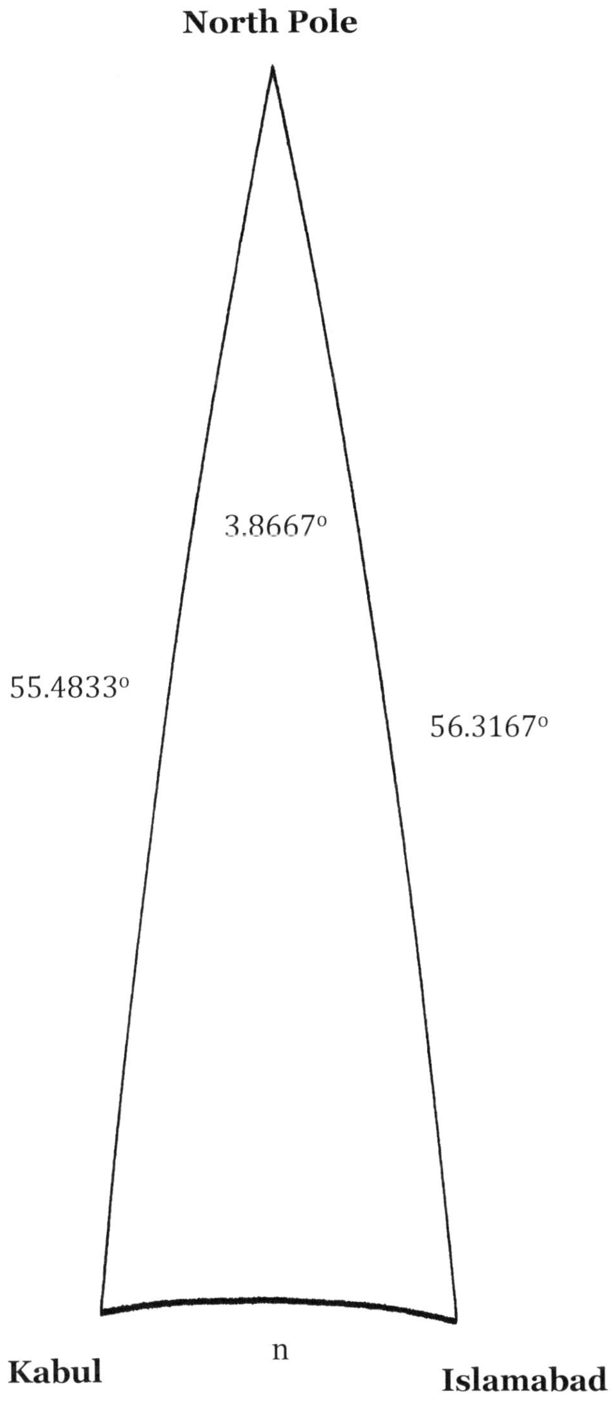
North Pole
3.8667°
55.4833°
56.3167°
n
Kabul
Islamabad

Chapter 10

Problem 2

Solution

Male, Maldives is located at $4^o10' \text{ N} = \left(4 + {}^{10}/_{60}\right)^o \text{ N} \approx 4.1667^o \text{ N}$,

$73^o30' \text{ E} = \left(73 + {}^{30}/_{60}\right)^o \text{ E} = 73.5000^o \text{ E}$.

Dhaka, Bangladesh is located at $23^o43' \text{ N} = \left(23 + {}^{43}/_{60}\right)^o \text{ N} \approx 23.7167^o \text{ N}$,

$90^o24' \text{ E} = \left(90 + {}^{24}/_{60}\right)^o \text{ E} = 90.4000^o \text{ E}$.

Two sides of the spherical triangle are $90^o - 4.1667^o \approx 85.8333^o$ and $90^o - 23.7167^o \approx 66.2833^o$, respectively. The angle at the North Pole is $90.4000^o - 73.5000^o \approx 16.9000^o$.

See the illustration on the next page.

It follows that

$\cos n = \cos(85.8333^o)\cos(66.2833^o) + \sin(85.8333^o)\ \sin(66.2833^o)\cos(16.9000^o)$

$\cos n \approx 0.902915213227 \Rightarrow n \approx \cos^{-1}(0.902915213227) \Rightarrow n \approx 25.4560551517^o$.

Therefore, the approximate distance between the two capital cities is

$(25.4560551517^o)\left({}^{\pi}/_{180^o}\right)(3959 \text{ miles})$

$\approx (0.444291976967)(3959 \text{ miles})$

$\approx$ 1,759 miles.

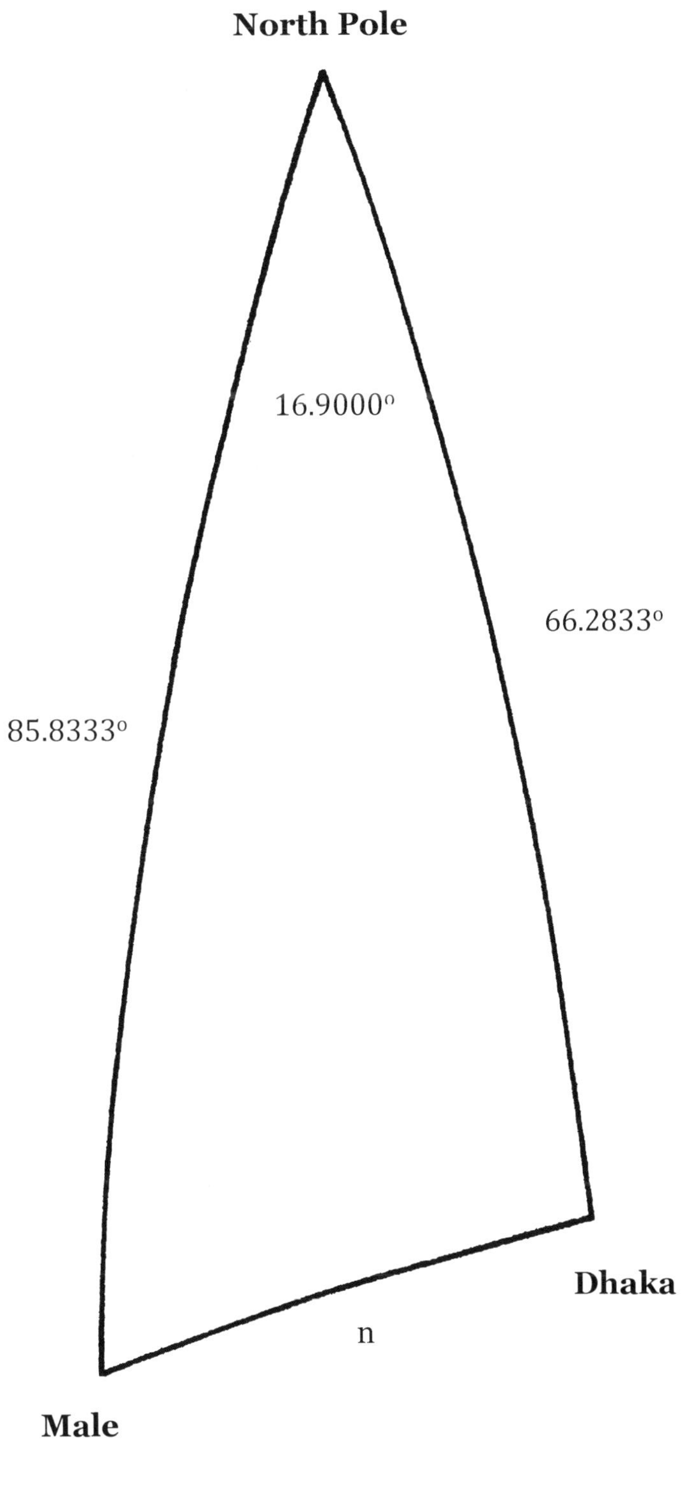
North Pole
16.9000°
66.2833°
85.8333°
Dhaka
n
Male

Chapter 10

Problem 3

Solution

New Delhi, India is located at $28^o36'$ N $= \left(28 + {}^{36}/_{60}\right)^o$ N $= 28.6000^o$ N,

$77^o12'$ E $= \left(77 + {}^{12}/_{60}\right)^o$ E $\approx 77.2000^o$ E.

Kathmandu, Nepal is located at $27^o43'$ N $= \left(27 + {}^{43}/_{60}\right)^o$ N $\approx 27.7167^o$ N,

$85^o19'$ E $= \left(85 + {}^{19}/_{60}\right)^o$ E $\approx 85.3167^o$ E.

Two sides of the spherical triangle are $90^o - 28.6000^o = 61.4000^o$ and $90^o - 27.7167^o = 62.2833^o$, respectively. The angle at the North Pole is $85.3167^o - 77.2000^o = 8.1167^o$.

See the illustration on the next page.

It follows that

$\cos n = \cos(61.4000^o)\cos(62.2833^o) + \sin(61.4000^o)\ \sin(62.2833^o)\cos(8.1167^o)$

$\cos n \approx 0.992095187755 \ \Rightarrow\ n \approx \cos^{-1}(0.992095187755) \ \Rightarrow\ n \approx 7.20891508558^o$.

Therefore, the approximate distance between the two capital cities is

$(7.20891508558^o)\left({}^{\pi}/_{180^o}\right)(3959 \text{ miles})$

$\approx (0.12581930374)(3959 \text{ miles})$

$\approx\ 498\ $ miles.

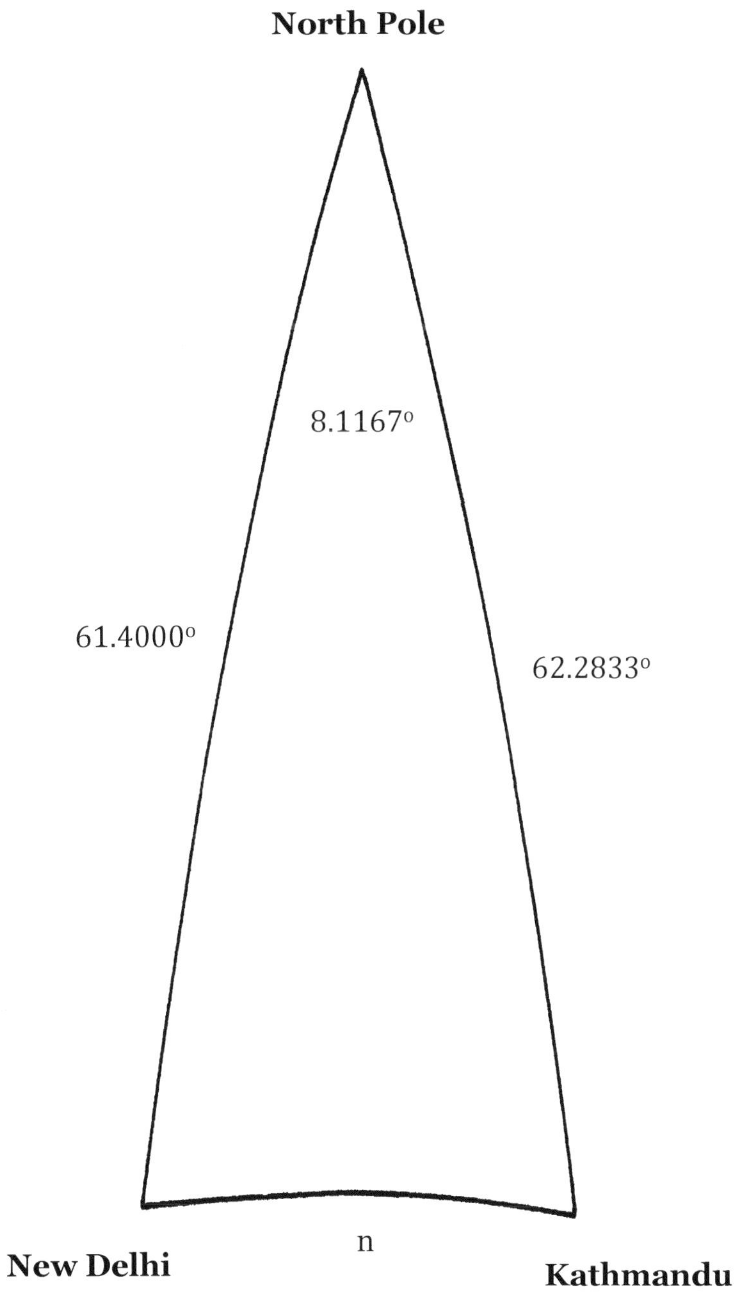
North Pole
8.1167°
61.4000°
62.2833°
n
New Delhi
Kathmandu

Chapter 11

Southeast Asia

Example 1

Beijing, China is located at $39^o55' \text{ N} = \left(39 + {}^{55}/_{60}\right)^o \text{ N} \approx 39.9167^o \text{ N}$,

$116^o23' \text{ E} = \left(116 + {}^{23}/_{60}\right)^o \text{ E} \approx 116.3833^o \text{ E}$.

Tokyo, Japan is located at $35^o41' \text{ N} = \left(35 + {}^{41}/_{60}\right)^o \text{ N} \approx 35.6833^o \text{ N}$,

$139^o45' \text{ E} = \left(139 + {}^{45}/_{60}\right)^o \text{ E} = 139.7500^o \text{ E}$.

Two sides of the spherical triangle are $90^o - 39.9167^o = 50.0833^o$ and $90^o - 35.6833^o = 54.3167^o$, respectively. The angle at the North Pole is $139.7500^o - 116.3833^o = 23.3667^o$.

See the illustration on the next page.

It follows that

$\cos \text{n} = \cos(50.0833^o) \cos(54.3167^o) + \sin(50.0833^o) \sin(54.3167^o) \cos(23.3667^o)$

$\cos \text{n} \approx 0.946178029251 \Rightarrow \text{n} \approx \cos^{-1}(0.946178029251) \Rightarrow \text{n} \approx 18.8836012386^o$.

Therefore, the approximate distance between the two capital cities is

$(18.8836012386^o)\left({}^{\pi}/_{180^o}\right)(3959 \text{ miles})$

$\approx (0.329581016247)(3959 \text{ miles})$

$\approx$ 1,305 miles.

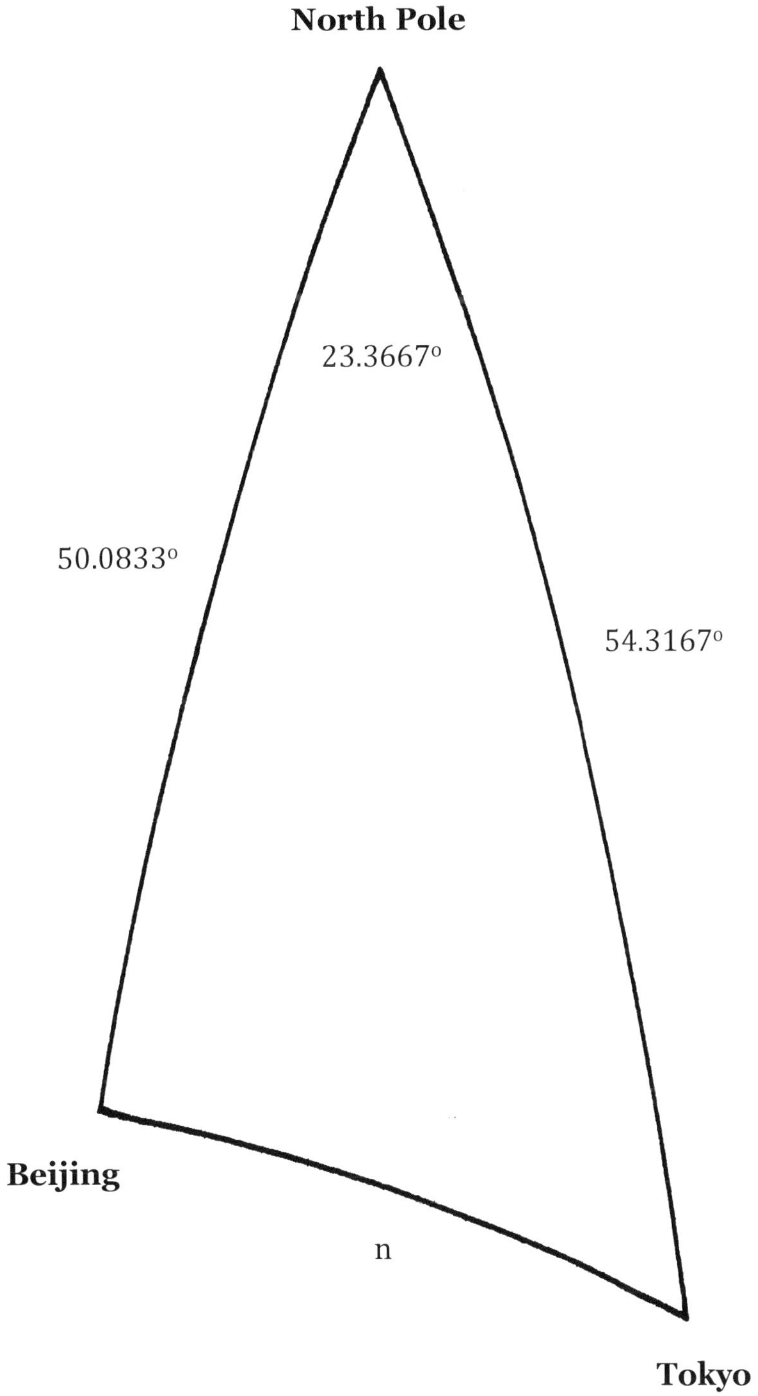
North Pole
23.3667°
50.0833°
54.3167°
Beijing
n
Tokyo

Example 2

Seoul, South Korea is located at $37^o 33' \text{ N} = \left(37 + {}^{33}/_{60}\right)^o \text{ N} = 37.5500^o \text{ N}$,
$126^o 59' \text{ E} = \left(126 + {}^{59}/_{60}\right)^o \text{ E} \approx 126.9833^o \text{ E}$.

Port Moresby, Papua New Guinea is located at

$9^o 27' \text{ S} = \left(9 + {}^{27}/_{60}\right)^o \text{ S} = 9.4500^o \text{ S}$, $147^o 11' \text{ E} = \left(147 + {}^{11}/_{60}\right)^o \text{ E} \approx 147.1833^o \text{ E}$.

Two sides of the spherical triangle are $90^o - 37.5500^o = 52.4500^o$ and $90^o + 9.4500^o = 99.4500^o$, respectively. The angle at the North Pole is $147.1833^o - 126.9833^o = 20.2000^o$.

See the illustration on the next page.

It follows that

$$\cos n = \cos(52.4500^o)\cos(99.4500^o) + \sin(52.4500^o)\sin(99.4500^o)\cos(20.2000^o)$$

$$\cos n \approx 0.633896052722 \Rightarrow n \approx \cos^{-1}(0.633896052722) \Rightarrow n \approx 50.6618452318^o.$$

Therefore, the approximate distance between the two capital cities is
$(50.6618452318^o)\left({}^{\pi}/_{180^o}\right)(3959 \text{ miles})$

$\approx (0.884216004431)(3959 \text{ miles})$

$\approx$ 3,501 miles.

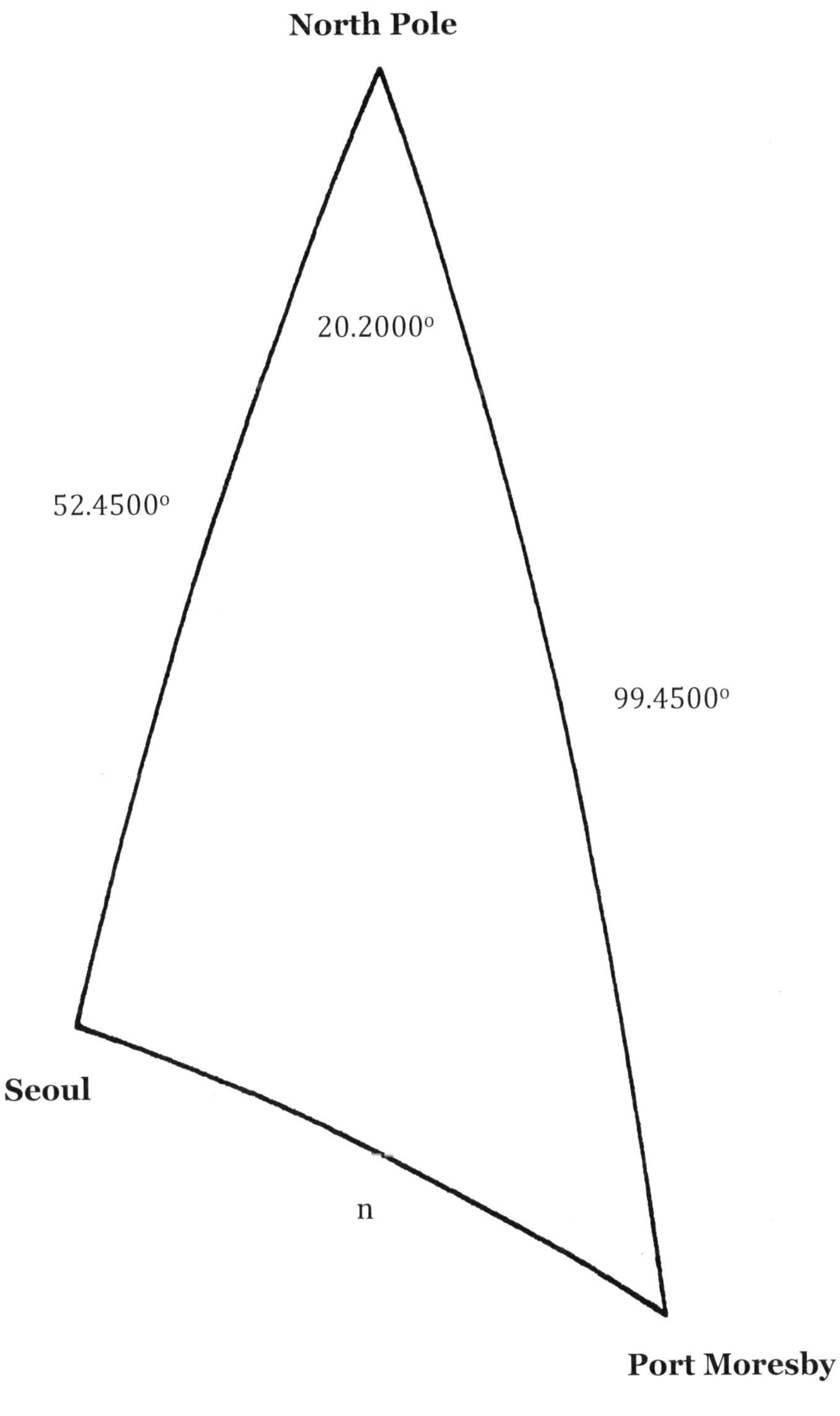
North Pole
20.2000°
52.4500°
99.4500°
Seoul
n
Port Moresby

Chapter 11 Problems:

Approximate the distance between the following capitals in Southeast Asia.

1. Bandar Seri Begawan, Brunei ($4^{o}53'$ N $114^{o}56'$ E)
   to Dili, Timor-Leste ($8^{o}35'$ S $125^{o}36'$ E)

2. Rangoon, Burma ($16^{o}48'$ N $96^{o}9'$ E)
   to Kuala Lumpur, Malaysia ($3^{o}10'$ N $101^{o}42'$ E)

3. Phnom Penh, Cambodia ($11^{o}33'$ N $104^{o}55'$ E)
   to Manila, Philippines ($14^{o}36'$ N $120^{o}58'$ E)

4. Vientiane, Laos($17^{o}58'$ N$102^{o}36'$ E)
   to Hanoi, Vietnam($21^{o}2'$ N$105^{o}51'$ E)

5. Ulaanbaatar, Mongolia ($47^{o}55'$ N $106^{o}55'$ E)
   to Taipei, Taiwan ($25^{o}2'$ N $121^{o}31'$ E)

6. Singapore, Singapore ($1^{o}17'$ N $103^{o}51'$ E)
   to Pyongyang, North Korea ($39^{o}1'$ N $125^{o}45'$ E)

7. Bangkok, Thailand ($13^{o}45'$ N $100^{o}31'$ E)
   to Jakarta, Indonesia ($6^{o}10'$ S $106^{o}49'$ E)

Chapter 11

Problem 1

Solution

Bandar Seri Begawan, Brunei is located at $4^o53'\text{ N} = \left(4 + {}^{53}/_{60}\right)^o \text{ N} \approx 4.8833^o \text{ N}$,

$114^o56'\text{ E} = \left(114 + {}^{56}/_{60}\right)^o \text{ E} \approx 114.9333^o \text{ E}$.

Dili, Timor-Leste is located at $8^o35'\text{ S} = \left(8 + {}^{35}/_{60}\right)^o \text{ S} \approx 8.5833^o \text{ S}$,

$125^o36'\text{ E} = \left(125 + {}^{36}/_{60}\right)^o \text{ E} = 125.6000^o \text{ E}$.

Two sides of the spherical triangle are $90^o - 4.8833^o = 85.1167^o$ and $90^o + 8.5833^o = 98.5833^o$, respectively. The angle at the North Pole is $125.6000^o - 114.9333^o = 10.6667^o$.

See the illustration on the next page.

It follows that

$\cos n = \cos(85.1167^o)\cos(98.5833^o) + \sin(85.1167^o)\sin(98.5833^o)\cos(10.6667^o)$

$\cos n \approx 0.955481929461 \;\Rightarrow\; n \approx \cos^{-1}(0.955481929461) \;\Rightarrow\; n \approx 17.1605132549^o$.

Therefore, the approximate distance between the two capital cities is

$(17.1605132549^o)\left({}^{\pi}/_{180^o}\right)(3959 \text{ miles})$

$\approx (0.29950745763)(3959 \text{ miles})$

$\approx 1{,}186$ miles.

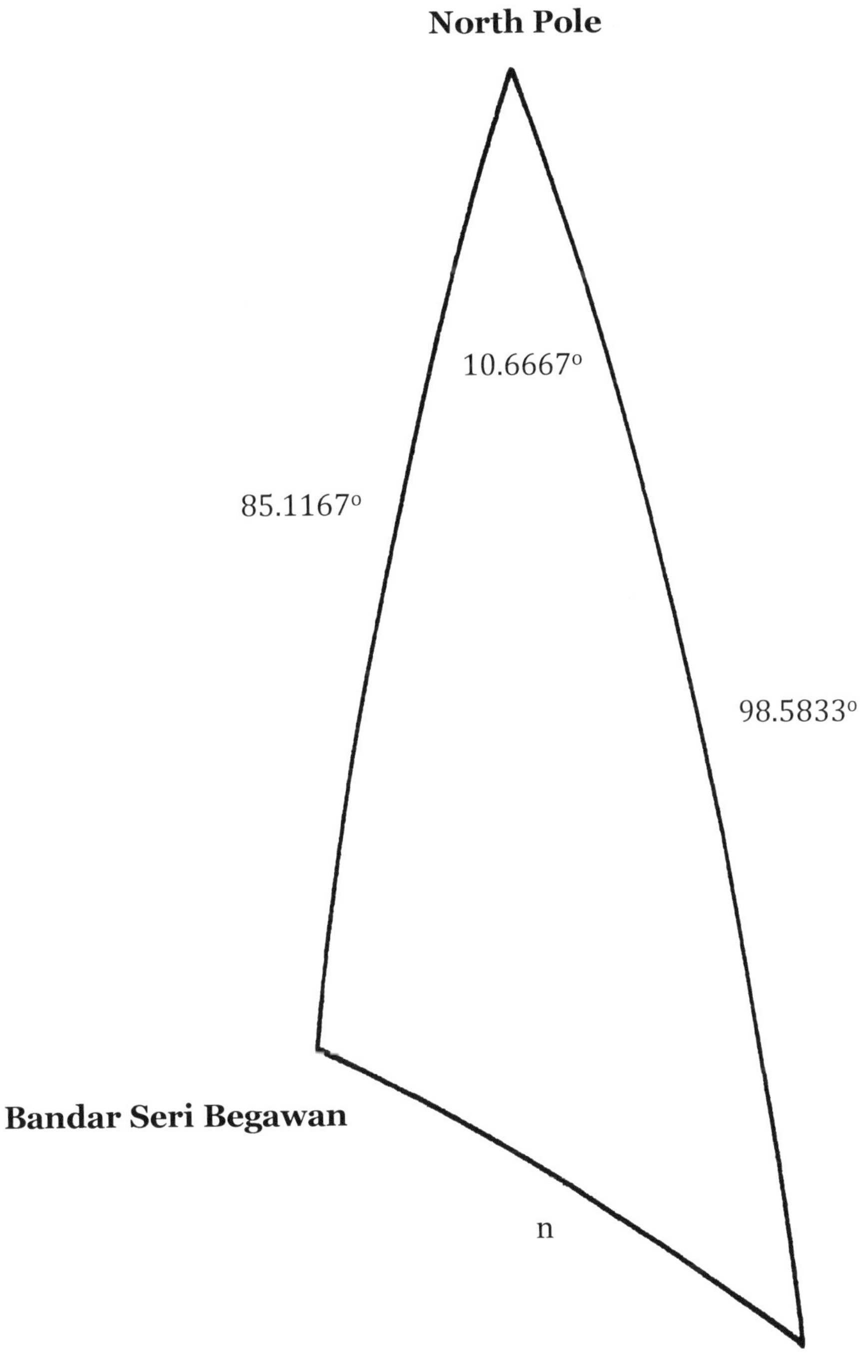
North Pole
10.6667°
85.1167°
98.5833°
Bandar Seri Begawan
n
Dili

Chapter 11
Problem 2
Solution

Rangoon, Burma is located at $16^o48' \text{ N} = \left(16 + {}^{48}/_{60}\right)^o \text{ N} = 16.8000^o \text{ N}$,
$96^o9' \text{ E} = \left(96 + {}^{9}/_{60}\right)^o \text{ E} = 96.1500^o \text{ E}$.

Kuala Lumpur, Malaysia is located at $3^o10' \text{ N} = \left(3 + {}^{10}/_{60}\right)^o \text{ N} \approx 3.1667^o \text{ N}$,
$101^o42' \text{ E} = \left(101 + {}^{42}/_{60}\right)^o \text{ E} = 101.7000^o \text{ E}$.

Two sides of the spherical triangle are $90^o - 16.8000^o = 73.2000^o$ and
$90^o - 3.1667^o = 86.8333$, respectively. The angle at the North Pole is
$101.7000^o - 96.1500^o = 5.5500^o$.

See the illustration on the next page.

It follows that
$$\cos \text{n} = \cos(73.2000^o)\cos(86.8333^o) + \sin(73.2000^o)\ \sin(86.8333^o)\cos(5.5500^o)$$
$$\cos \text{n} \approx 0.967343282455\ \Rightarrow\ \text{n} \approx \cos^{-1}(0.967343282455)\ \Rightarrow\ \text{n} \approx 14.6829365357^o.$$

Therefore, the approximate distance between the two capital cities is
$$(14.6829365357^o)\left({}^{\pi}/_{180^o}\right)(3959 \text{ miles})$$
$$\approx (0.25626558641)(3959 \text{ miles})$$
$$\approx 1{,}015 \text{ miles}.$$

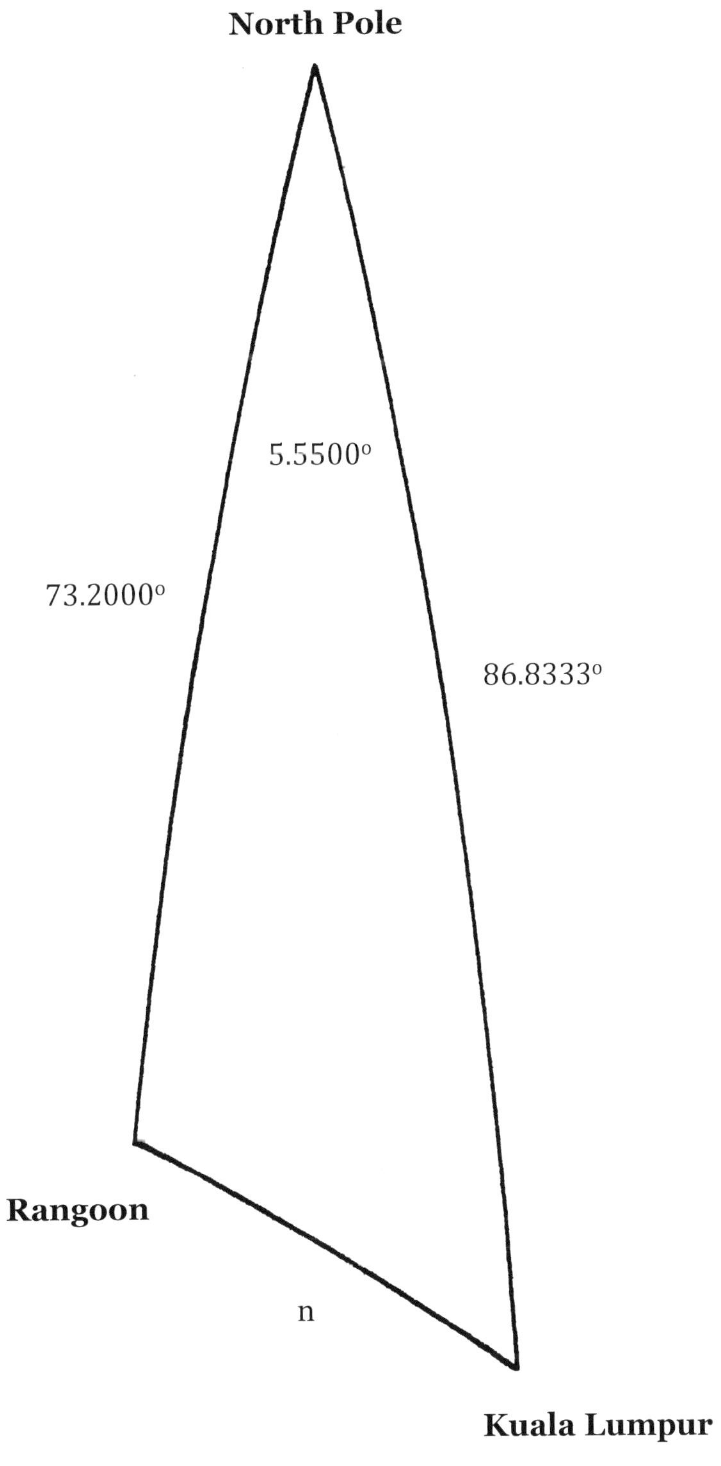
North Pole
5.5500°
73.2000°
86.8333°
Rangoon
n
Kuala Lumpur

Chapter 11
Problem 3
Solution

Phnom Penh, Cambodia is located at $11^o 33'\ \mathrm{N} = \left(11 + {}^{33}/_{60}\right)^o\ \mathrm{N} = 11.5500^o\ \mathrm{N}$,

$104^o 55'\ \mathrm{E} = \left(104 + {}^{55}/_{60}\right)^o\ \mathrm{E} \approx 104.9167^o\ \mathrm{E}$.

Manila, Philippines is located at $14^o 36'\ \mathrm{N} = \left(14 + {}^{36}/_{60}\right)^o\ \mathrm{N} = 14.6000^o\ \mathrm{N}$,

$120^o 58'\ \mathrm{E} = \left(120 + {}^{58}/_{60}\right)^o\ \mathrm{E} \approx 120.9667^o\ \mathrm{E}$.

Two sides of the spherical triangle are $90^o - 11.5500^o = 78.4500^o$ and $90^o - 14.6000^o = 75.4000^o$, respectively. The angle at the North Pole is $120.9667^o - 104.9167^o = 16.0500^o$.

See the illustration on the next page.

It follows that

$\cos n = \cos(78.4500^o)\cos(75.4000^o) + \sin(78.4500^o)\ \sin(75.4000^o)\cos(16.0500^o)$

$\cos n \approx 0.96162677243 \Rightarrow n \approx \cos^{-1}(0.96162677243) \Rightarrow n \approx 15.9239367194^o$.

Therefore, the approximate distance between the two capital cities is

$(15.9239367194^o)\left({}^{\pi}/_{180^o}\right)(3959\ \text{miles})$

$\approx (0.277925125632)(3959\ \text{miles})$

$\approx\ 1{,}100\ \text{miles}$.

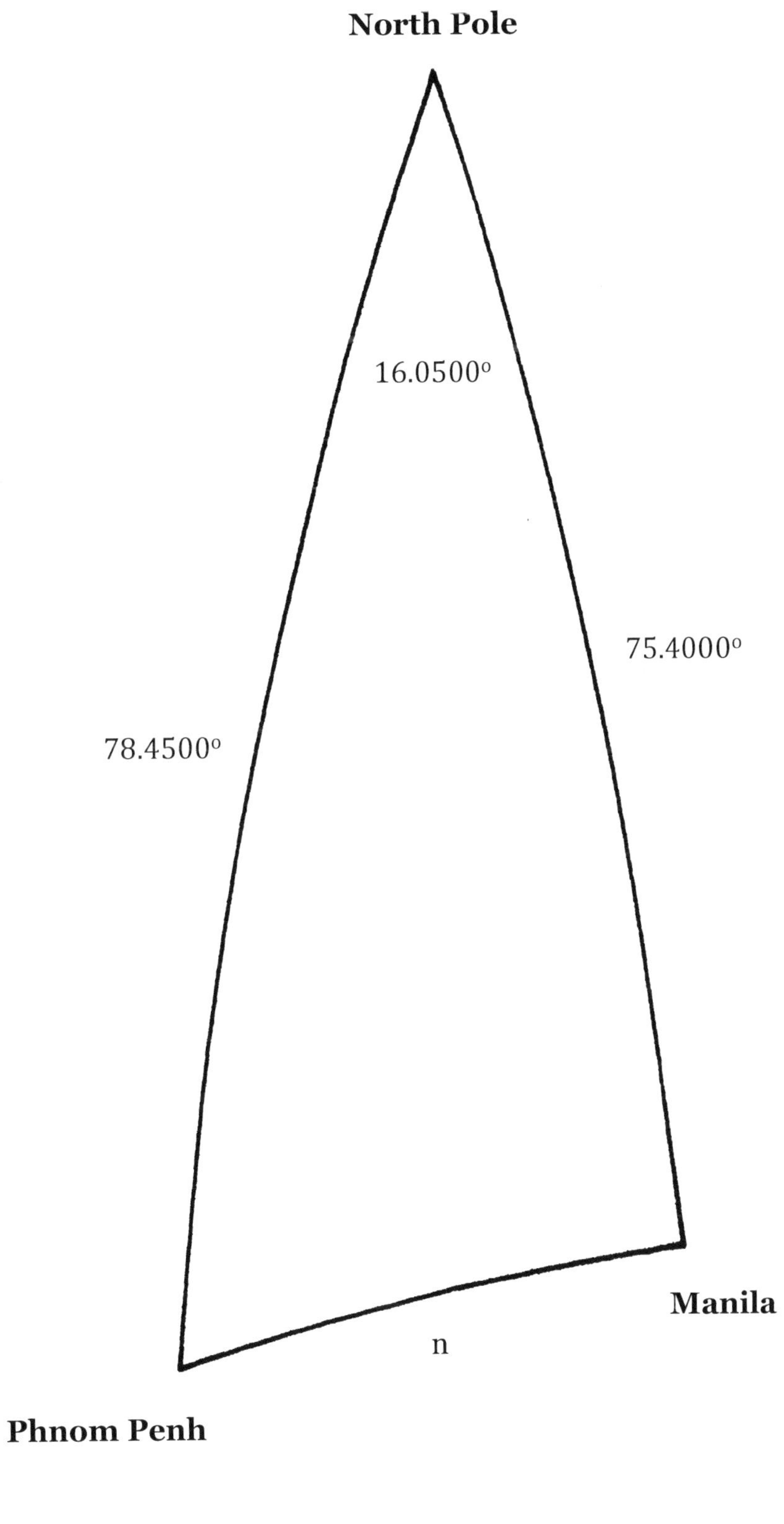
North Pole
16.0500°
75.4000°
78.4500°
Manila
n
Phnom Penh

Chapter 11

Problem 4

Solution

Vientiane, Laos is located at $17^o58'\ \mathrm{N} = \left(17 + {}^{58}/_{60}\right)^o\ \mathrm{N} \approx 17.9667^o\ \mathrm{N}$, $102^o36'\ \mathrm{E} = \left(102 + {}^{36}/_{60}\right)^o\ \mathrm{E} = 102.6000^o\ \mathrm{E}$.

Hanoi, Vietnam is located at $21^o2'\ \mathrm{N} = \left(21 + {}^{2}/_{60}\right)^o\ \mathrm{N} \approx 21.0333^o\ \mathrm{N}$, $105^o51'\ \mathrm{E} = \left(105 + {}^{51}/_{60}\right)^o\ \mathrm{E} = 105.8500^o\ \mathrm{E}$.

Two sides of the spherical triangle are $90^o - 17.9667^o = 72.0333^o$ and $90^o - 21.0333^o = 68.9667^o$, respectively. The angle at the North Pole is $105.8500^o - 102.6000^o = 3.2500^o$.

See the illustration on the next page.

It follows that

$$\cos n = \cos(72.0333^o)\cos(68.9667^o) + \sin(72.0333^o)\ \sin(68.9667^o)\cos(3.2500^o)$$

$$\cos n \approx 0.997140058822 \ \Rightarrow\ n \approx \cos^{-1}(0.997140058822) \ \Rightarrow\ n \approx 4.33430770317^o.$$

Therefore, the approximate distance between the two capital cities is

$$(4.33430770317^o)\left({}^{\pi}/_{180^o}\right)(3959 \text{ miles})$$

$$\approx (0.075647940215)(3959 \text{ miles})$$

$$\approx 299 \text{ miles}.$$

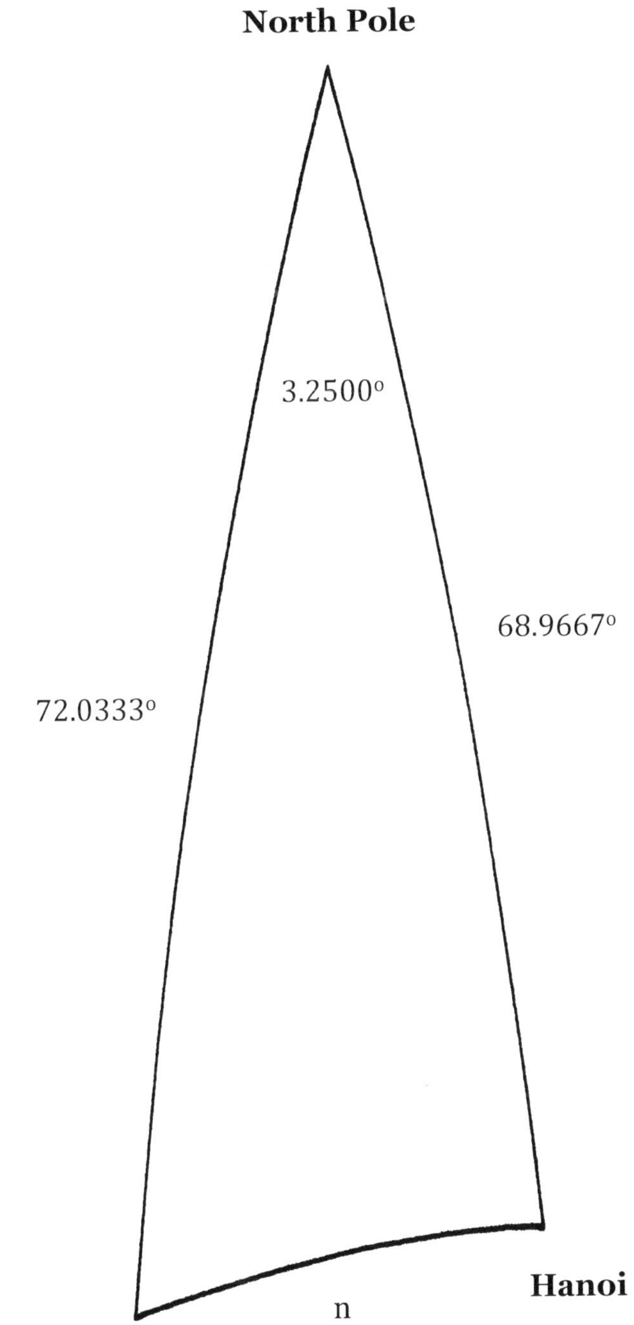
North Pole
3.2500°
68.9667°
72.0333°
Hanoi
n
Vientiane

Chapter 11
Problem 5
Solution

Ulaanbaatar, Mongolia is located at $47^o55'\ \mathrm{N} = \left(47 + {}^{55}\!/_{60}\right)^o\ \mathrm{N} \approx 47.9167^o\ \mathrm{N}$,

$106^o55'\ \mathrm{E} = \left(106 + {}^{55}\!/_{60}\right)^o\ \mathrm{E} \approx 106.9167^o\ \mathrm{E}$.

Taipei, Taiwan is located at $25^o2'\ \mathrm{N} = \left(25 + {}^{2}\!/_{60}\right)^o\ \mathrm{N} \approx 25.0333^o\ \mathrm{N}$,

$121^o31'\ \mathrm{E} = \left(121 + {}^{31}\!/_{60}\right)^o\ \mathrm{E} \approx 121.5167^o\ \mathrm{E}$.

Two sides of the spherical triangle are $90^o - 47.9167^o = 42.0833^o$ and $90^o - 25.0333^o = 64.9667^o$, respectively. The angle at the North Pole is $121.5167^o - 106.9167^o = 14.6000^o$.

See the illustration on the next page.

It follows that

$\cos \mathrm{n} = \cos(42.0833^o)\cos(64.9667^o) + \sin(42.0833^o)\ \sin(64.9667^o)\cos(14.6000^o)$

$\cos \mathrm{n} \approx 0.901689431147 \Rightarrow \mathrm{n} \approx \cos^{-1}(0.901689431147) \Rightarrow \mathrm{n} \approx 25.6189682788^o$.

Therefore, the approximate distance between the two capital cities is

$(25.6189682788^o)\left({}^{\pi}\!/_{180^o}\right)(3959\ \text{miles})$

$\approx (0.447135347428)(3959\ \text{miles})$

$\approx 1{,}770\ \text{miles}$.

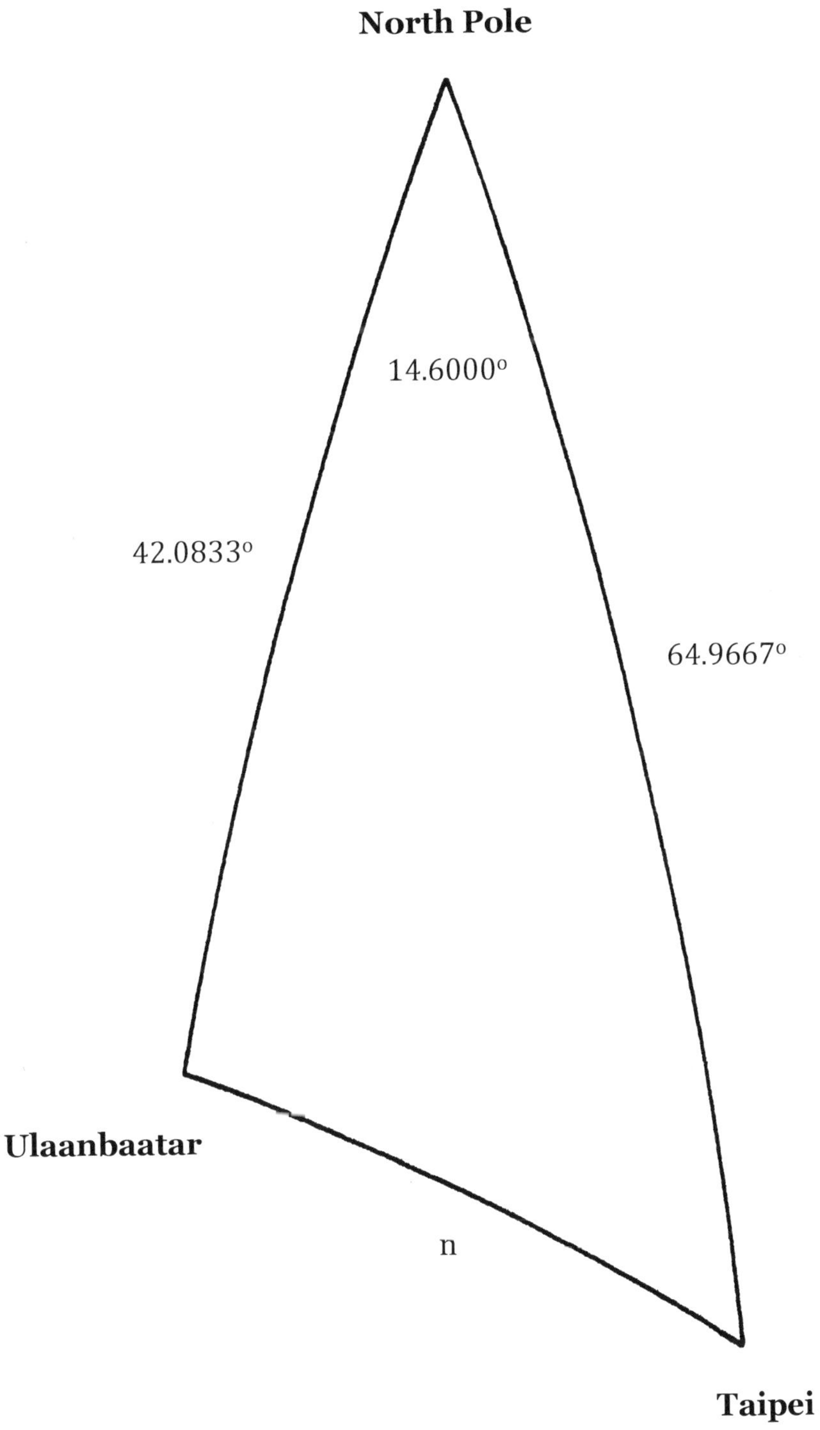
North Pole
14.6000°
42.0833°
64.9667°
Ulaanbaatar
n
Taipei

Chapter 11
Problem 6
Solution

Singapore, Singapore is located at $1^o17'$ N $= \left(1 + {}^{17}/_{60}\right)^o$ N $\approx 1.2833^o$ N ,
$103^o51'$ E $= \left(103 + {}^{51}/_{60}\right)^o$ E $= 103.8500^o$ E .

Pyongyang, North Korea is located at $39^o1'$ N $= \left(39 + {}^{1}/_{60}\right)^o$ N $\approx 39.0167^o$ N ,
$125^o45'$ E $= \left(125 + {}^{45}/_{60}\right)^o$ E $= 125.7500^o$ E .

Two sides of the spherical triangle are $90^o - 1.2833^o = 88.7167^o$ and
$90^o - 39.0167^o = 50.9833^o$, respectively. The angle at the North Pole is
$125.7500^o - 103.8500^o = 21.9000^o$.

See the illustration on the next page.

It follows that

$\cos n = \cos(88.7167^o)\cos(50.9833^o) + \sin(88.7167^o)\ \sin(50.9833^o)\cos(21.9000^o)$

$\cos n \approx 0.734812453518 \Rightarrow n \approx \cos^{-1}(0.734812453518) \Rightarrow n \approx 42.7086289515^o$.

Therefore, the approximate distance between the two capital cities is

$(42.7086289515^o)\left({}^{\pi}/_{180^o}\right)(3959 \text{ miles})$

$\approx (0.745406194216)(3959 \text{ miles})$

$\approx 2{,}951$ miles.

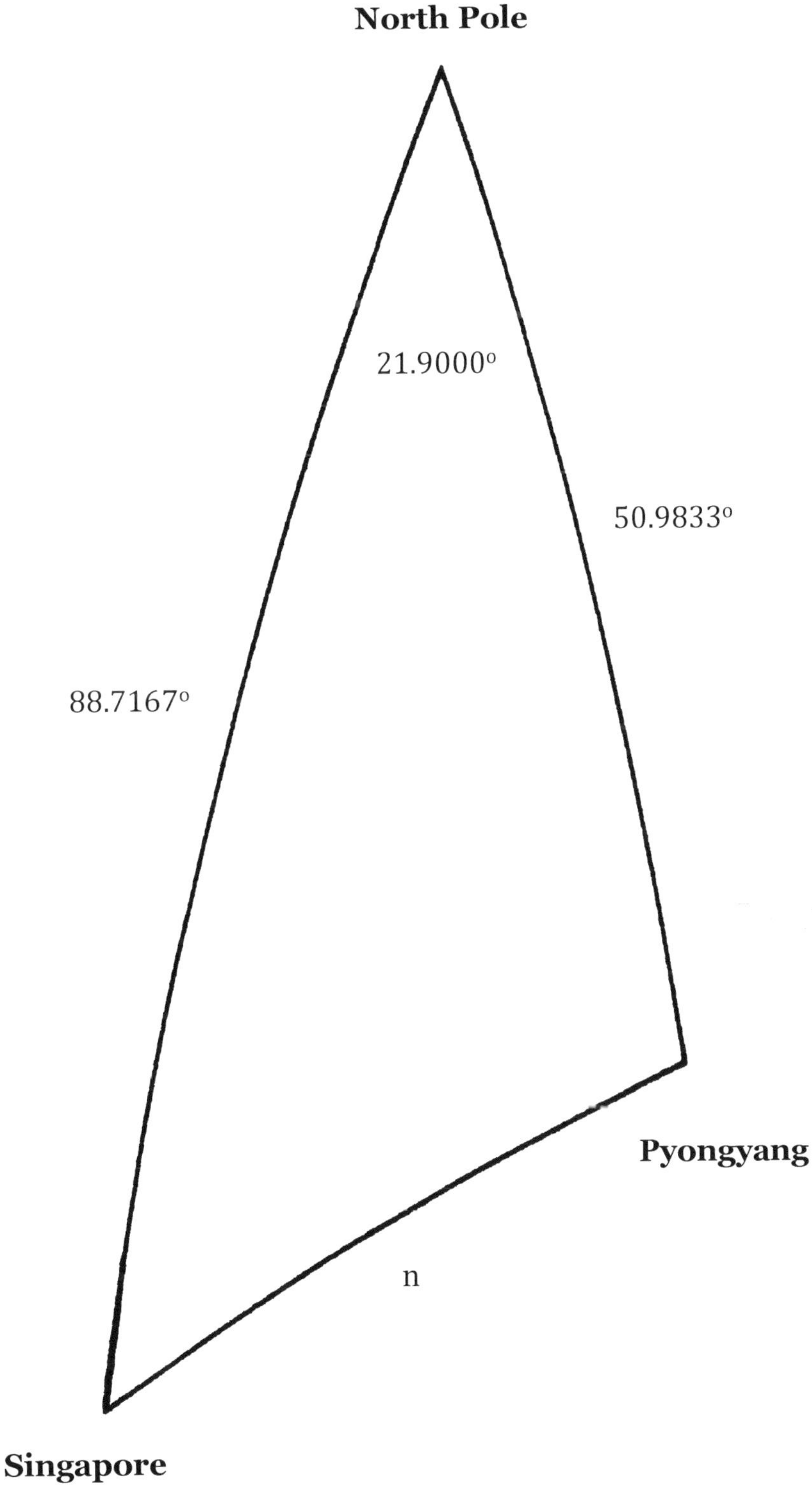
North Pole
21.9000°
50.9833°
88.7167°
Pyongyang
n
Singapore

Chapter 11
Problem 7
Solution

Bangkok, Thailand is located at $13^{o}45'$ N $= \left(13 + {}^{45}/_{60}\right)^{o}$ N $= 13.7500^{o}$ N, $100^{o}31'$ E $= \left(100 + {}^{31}/_{60}\right)^{o}$ E $\approx 100.5167^{o}$ E.

Jakarta, Indonesia is located at $6^{o}10'$ S $= \left(6 + {}^{10}/_{60}\right)^{o}$ S $\approx 6.1667^{o}$ S,

$106^{o}49'$ E $= \left(106 + {}^{49}/_{60}\right)^{o}$ E $\approx 106.8167^{o}$ E.

Two sides of the spherical triangle are $90^{o} - 13.7500^{o} = 76.2500^{o}$ and $90^{o} + 6.1667^{o} = 96.1667^{o}$, respectively. The angle at the North Pole is $106.8167^{o} - 100.5167^{o} = 6.3000^{o}$.

See the illustration on the next page.

It follows that

$\cos n = \cos(76.2500^{o})\cos(96.1667^{o}) + \sin(76.2500^{o})\sin(96.1667^{o})\cos(6.3000^{o})$

$\cos n \approx 0.934356841723 \Rightarrow n \approx \cos^{-1}(0.934356841723) \Rightarrow n \approx 20.8755145315^{o}$.

Therefore, the approximate distance between the two capital cities is

$(20.8755145315^{o})\left({}^{\pi}/_{180^{o}}\right)(3959 \text{ miles})$

$\approx (0.364346461623)(3959 \text{ miles})$

$\approx 1{,}442$ miles.

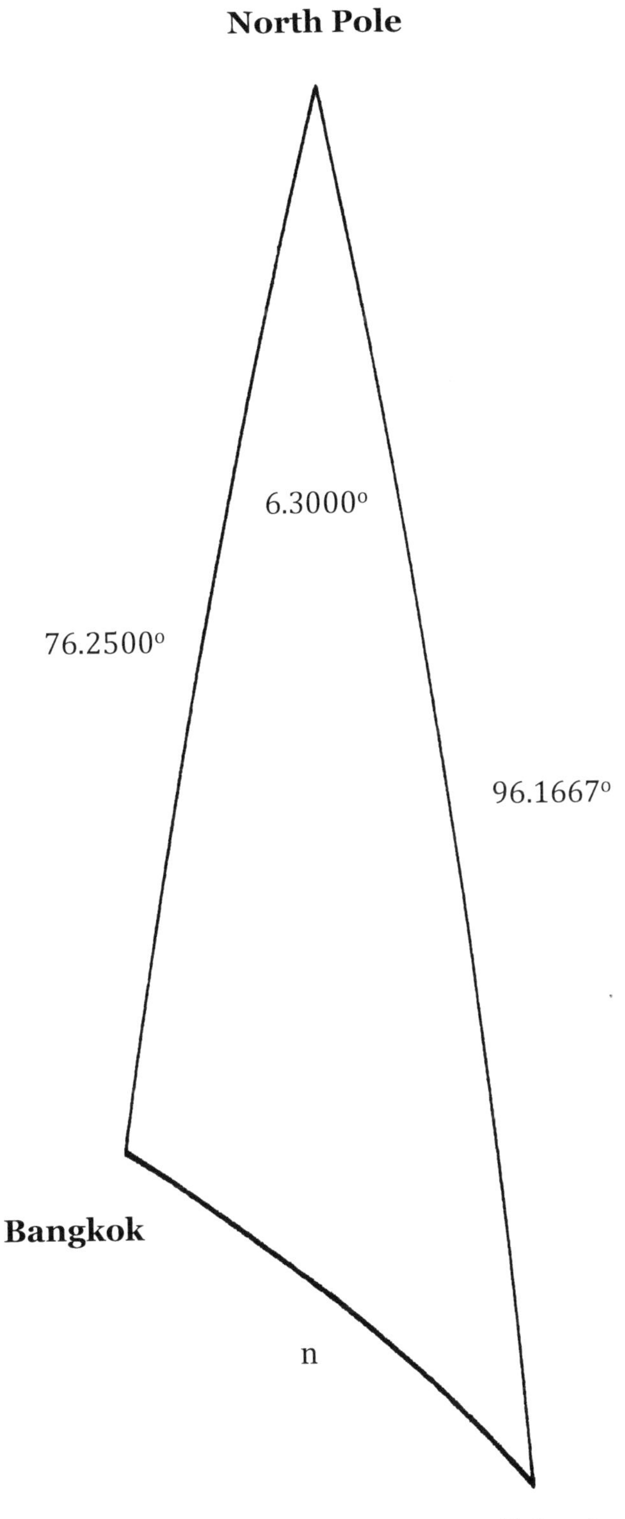
North Pole
6.3000°
76.2500°
96.1667°
Bangkok
n
Jakarta

References

Spherical law of cosines. (2017, June 24). Retrieved June 25, 2017, from https://en.wikipedia.org/wiki/Spherical_law_of_cosines

Contact CIA. (2016, April 01). Retrieved June 25, 2017, from https://www.cia.gov/library/publications/resources/the-world-factbook/

Kells, L. M., Kern, W. F., & Bland, J. R. (1942). Spherical trigonometry, with naval and military applications. New York: McGraw-Hill Book Co.

Index of Capitals, Countries and Continents

A

B

C

D

E

F

G

H

I

J

K

L

M

N

T

U

V

W

Y

Z

Made in United States
Orlando, FL
04 November 2026

90202178R00166